ROCK FORMATIONS AND UNUSUAL GEOLOGIC STRUCTURES

ROCK FORMATIONS AND UNUSUAL GEOLOGIC STRUCTURES

EXPLORING THE EARTH'S SURFACE

The Changing Earth Series

JON ERICKSON

Facts On File

ROCK FORMATIONS AND UNUSUAL GEOLOGIC STRUCTURES:
EXPLORING THE EARTH'S SURFACE

Facts On File, Inc.
460 Park Avenue South
New York NY 10016
USA

Library of Congress Cataloging-in-Publication Data
Erickson, Jon, 1948–
Rock formations and unusual geologic structures : exploring the earth's surface / Jon Erickson.
p. cm. — (The Changing earth series)
Includes bibliographical references and index.
ISBN 0-8160-2589-4 (alk. paper)
1. Geology. 2. Geomorphology. I. Title. II. Series: Erickson, Jon, 1948– Changing earth.
QE33.E75 1993
550—dc20 92-32097

A British CIP catalogue record for this book is available from the British Library.

Facts On File books are available at special discounts when purchased in bulk quantities for businesses, associations, institutions or sales promotions. Please contact our Special Sales Department in New York at 212/683-2244 or 800/322-8755.

Text design by Ron Monteleone
Jacket design by Catherine Hyman
Composition by Facts On File, Inc./Robert Yaffe
Manufactured by R. R. Donnelley & Sons, Inc.
Printed in the United States of America

10 9 8 7 6 5 4 3 2 1

This book is printed on acid-free paper.

CONTENTS

TABLES IN ROCK FORMATIONS AND UNUSUAL GEOLOGIC STRUCTURES

ACKNOWLEDGMENTS

The author would like to thank the following organizations for providing photographs for this book: the National Aeronautics and Space Administration (NASA), the National Park Service, the U.S. Air Force, the U.S. Department of Agriculture–Forest Service, the U.S. Department of Agriculture–Soil Conservation Service, the U.S. Geological Survey (USGS) and the U.S. Navy.

INTRODUCTION

The most impressive features on Earth are its geologic structures created by the interaction of movable plates, volcanic and magmatic activity, and deposition and erosion. The Earth is unique by being the only planet known to possess continents, which are essentially thick slabs of granite riding on a sea of molten rock in the upper mantle. The interaction of over a dozen crustal plates is responsible for all geologic activity taking place on the surface of the Earth. The constant shifting of the plates continues to generate new crust at spreading centers while destroying old crust at subduction zones. This activity makes the Earth a living planet in both the physical and biological senses.

No other terrain can compare with mountain ranges, which were created by the forces of uplift and erosion. The cores of mountain ranges contain some of the oldest rocks, which were thrust upward by powerful tectonic forces. The crust is also ripped by faults, which relieve pressures building up due to plate motions. Slippage along major fault systems is often accompanied by earthquakes, which can be very destructive as they wrench the landscape. A long belt of earthquakes wraps itself around the world and coincides with the boundaries between active plates. Earthquakes not only destroy entire cities but completely change the structure of the landscape, producing tall scarps and massive landslides.

A large portion of western North America contains a jumble of distinct rock units bounded by faults called terranes. They were skimmed off the ocean floor and plastered onto the leading edge as the continent drifted westward over the Pacific plate. Each block has geologic histories completely different from adjacent terranes and the adjoining continental masses. Many terranes traveled long distances, from as far away as the western Pacific. Over the past 200 million years, North America expanded by over 25 percent from the accretion of crustal fragments to its growing coast. The accreted terranes also played a major role in the creation of mountain ranges along convergent continental margins.

Most of the Earth's surface is covered by a thin veneer of sediments, which provide an impressive terrain of ragged mountains and jagged canyons. Alternating beds of sandstone, shale and limestone indicate different depositional environments as the coastline advanced and receded. The sedimentary environment is important for the formation of fossils, which provide insights into the history of the Earth. The constant shifting of sediments on the Earth's surface and the accumulation of deposits on the ocean floor ensure that the face of the Earth will continue to change.

Some of the most fascinating geologic features were carved out of the crust by erosion. The continents are delicately balanced by buoyancy and erosion, and annually some 25 million tons of sediment are carried off by rivers and dumped into the sea. Erosion has leveled the tallest mountains, gouged deep ravines and obliterated most geologic structures, including most man-made structures. No matter how pervasive is the formation of mountain ranges, they eventually lose the battle with erosion and are worn down to the level of the prevailing plain.

Much of the landscape in the northern latitudes owes its unusual topography to gigantic ice sheets that swept down from the polar regions during the last ice age. In some regions, the crust was scraped completely clean of sediments, exposing bare bedrock. In mountainous regions, the glaciers gouged out deep-sided valleys, marking their farthest extent by heaps of debris. In other areas, massive quantities of glacial sediments were deposited as the glaciers melted and retreated to the poles. These include elongated hillocks, sinuous sand deposits, circular depressions and immense boulder fields.

The surface of the Earth is sculpted by a number of forces, providing a cornucopia of unusual geologic structures, ranging from narrow spires, mesas, ragged crags and tall pillars carved out of stone; dikes and volcanic necks created by ancient volcanoes; arches and caves eroded out of solid rock; and meteorite and volcanic craters, just to name a few. These unique structures give our planet its highly diverse geology, without which the Earth would indeed be a dull place.

1

THE EARTH'S CRUST

The Earth, like all terrestrial planets, has a central core surrounded by an intermediate layer, or mantle, covered by a thin shell called the crust. The study of metallic meteorites, which once formed the cores of early planetoids that have since disintegrated, suggests that the Earth's core is composed of iron and nickel. The age of the Earth, estimated at 4.6 billion years, is based on an agreement between the ages of meteorites that are thought to have formed at the same time as the planet. There is also general agreement with the ages of lead isotopes in lunar rocks (Figure 1) as well as the Earth's oldest known rocks, which formed about 4.2 billion years ago, about the time the crust first segregated from the mantle.

In the Earth's early stages of formation, it was ceaselessly pounded by giant meteorite impacts, during which time the planet was struck by as many as three Mars-size bodies. One of these impactors might have been responsible for creating the moon by launching great quantities of material into Earth orbit, where it coalesced into a satellite, the largest in the Solar System with respect to its mother planet. The giant impacts also might have removed the Earth's pimordial atmosphere and initiated the formation of continents. The Earth is unique among planets because it is the only one known to have distinct continents (Figure 2), which are essentially thick slabs of granite riding on a sea of semimolten rock in the upper mantle.

Figure 1 A large lunar boulder field at Taurus-Littrow from *Apollo 17* in December 1972. Courtesy of NASA

This feature is responsible for making the Earth a living planet both physically and biologically.

THE PROTOEARTH

During the early developmental stages of the Solar System, a protoplanetary disk slowly revolved around the infant sun. It consisted of several bands of coarse particles called planetesimals, composed of primordial dust grains from a nearby supernova (Figure 3). The planetesimals stuck to each other by weak electrical and gravitational forces. They continued to grow by constantly colliding with each other as they swung around the sun in highly eccentric orbits. Perhaps as many as 100 trillion planetesimals orbited the sun during the first stages of development of the Solar System.

The planetesimals closest to the sun were composed mostly of stony and metallic minerals. They ranged in size from sand grains to large bodies more than 50 miles across, but most were pebble-size. The planetesimals forming farther away from the sun in the cold depths of space were composed mainly of frozen volatiles and gases. They coalesced into the large gaseous planets and the comets (Figure 4), which are composed of a mixture of rock and ices. The entire process of planetary formation was completed in less than half a million years.

The Earth and possibly its moon accreted into a homogeneous mixture of silicates and iron-nickel from numerous meteorite impacts. As the Earth grew, its orbit began to decay due to drag forces from leftover gases in interplanetary space. The formative planet slowly spiraled closer toward the sun, sweeping up more planetesimals on its way like a gravitational dust mop. There also might have been large planetesimals in orbit around the Earth as well as the moon. Eventually, the Earth's path around the sun was completely swept clean of planetesimals, and its orbit stabilized near where it is today.

High-intensity radioactive elements within the recently formed planet called radionuclides generated a tremendous amount of heat, melting the Earth from the inside out. This allowed the planet to segregate into concentric layers, with the heavier elements, mostly iron and nickel, falling

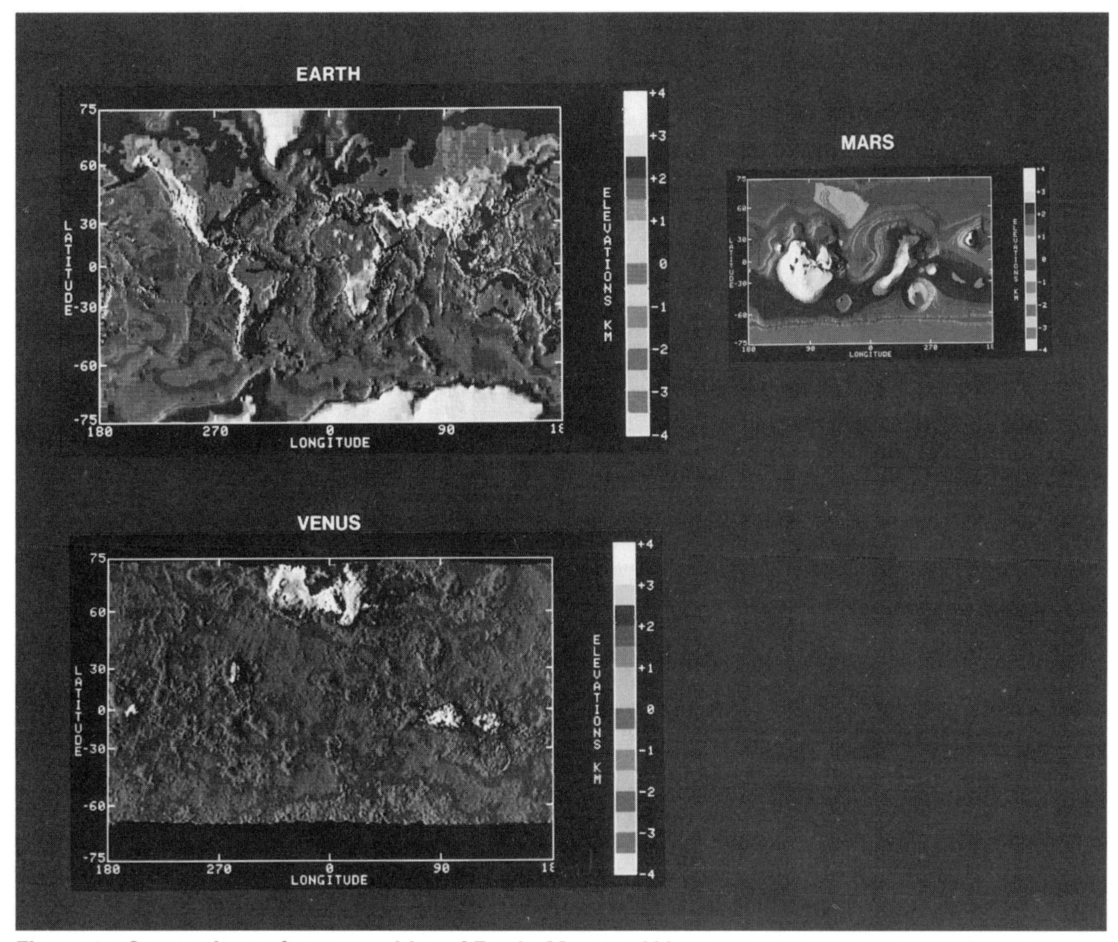

Figure 2 Comparison of topographies of Earth, Mars and Venus. Courtesy of NASA

inward toward the center under the pull of gravity and the lighter substances, mostly silicates, floating outward toward the surface due to their greater buoyancy. On its way downward, the molten iron scrubbed out and dragged down other elements called siderophiles, meaning iron lovers, such as platinum and iridium, which is why they are so rare in the Earth's crust. The entire process of segregation took only about 100 million years to complete, resulting in a planet that was partially molten throughout.

The presence of large amounts of leftover volatiles and gases that pervaded the early Solar System supplied the infant Earth with a dense atmosphere. Atmospheric pressures were over 100 times higher than they are today, creating surface temperatures greater than the melting point of rocks. A massive meteorite bombardment along with accompanying high

Figure 3 The Solar System condensed from a dense cloud similar to the Orion nebula. Courtesy of NASA

surface temperatures blasted away the Earth's original atmosphere, leaving the planet completely devoid of any gases, much like the moon is today.

Without an atmosphere to hold in the Earth's internally generated heat, the surface rapidly cooled, forming a primitive crust similar to the slag that forms on top of molten iron ore. This was not a true crust, however, because the interior of the Earth was still in a highly molten and agitated state, which kept the mantle well mixed and did not allow for chemical segregation of its various materials. Therefore, the density of rocks solidifying on the surface was not much different than that of the mantle, which made the crust highly unstable. The crust could have remelted on the surface, dove back into the mantle and remelted, or become top-heavy, overturned and remelted. Because the early crust was so unstable, there is no geologic record of the first half billion years of the Earth's existence.

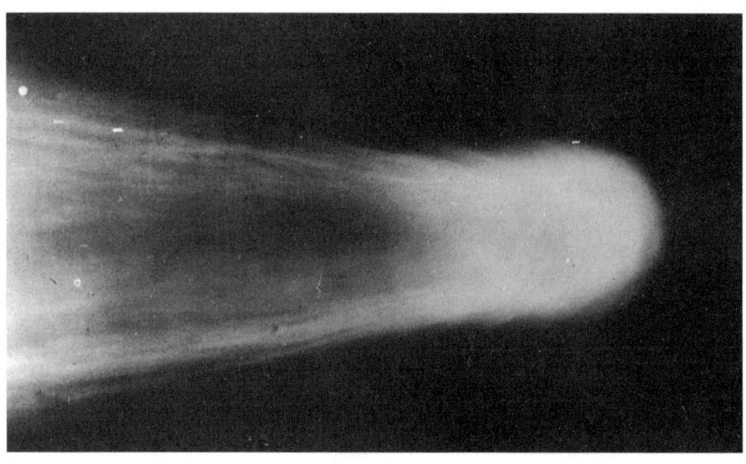

Figure 4 Halley's comet. Courtesy of NASA

As currents in the mantle began to slow down due to the loss of radiogenic heat supplied by the decay of radioactive elements, lighter rock materials migrated toward the surface, forming a basaltic crust like the film on top of cold pudding. Between 4.2 and 3.8 billion years ago, a massive meteorite shower bombarded the Earth and its moon. Near the end of this cratering period, several large meteorites

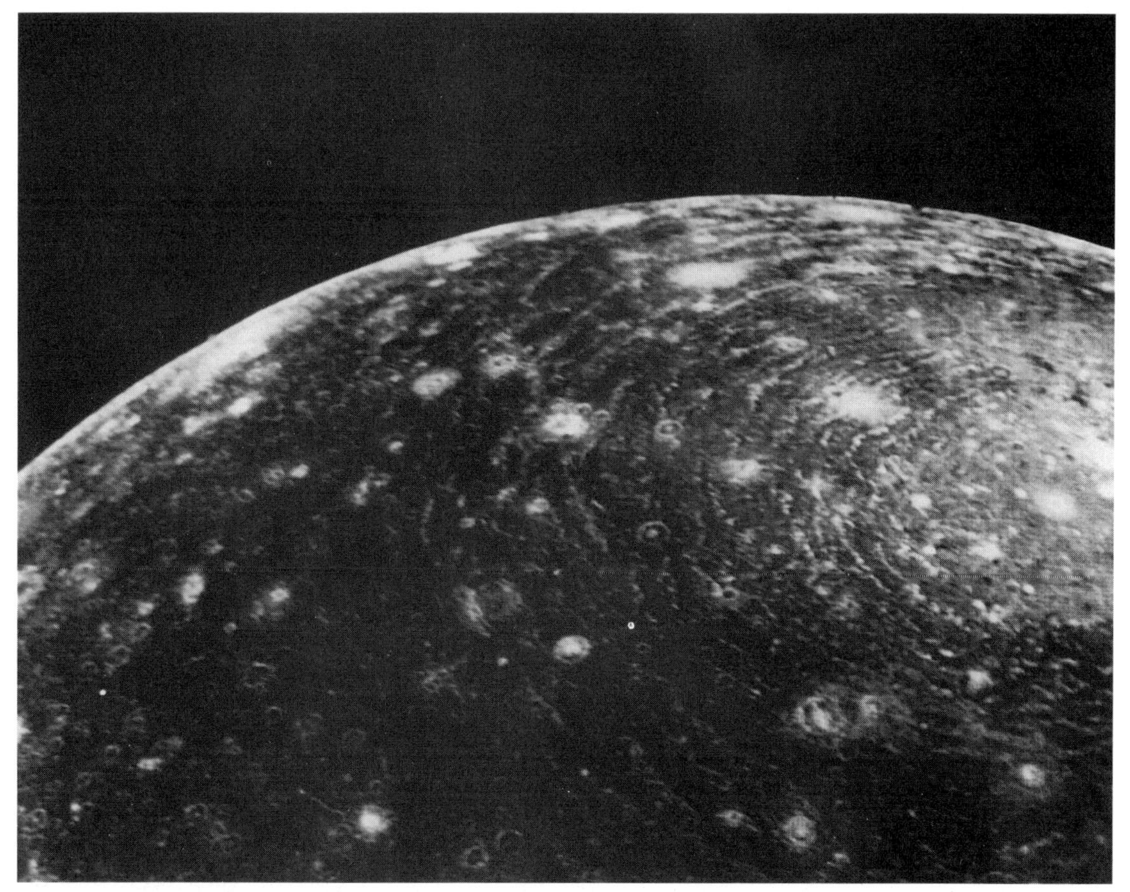

Figure 5 The prominent bull's-eye structure of Jupiter's moon Callisto appears to be a large impact basin in a crust composed of dirty ice. Courtesy of NASA

struck the lunar surface, breaking through the thin crust. Dark basalt rapidly escaped to the surface, flooding low-lying areas, which became wide lava plains called maria. The rest of the inner planets as well as the moons of the outer planets show numerous pockmarks from this invasion (Figure 5), and nothing of this magnitude has ever happened again.

By this time, the Earth had only a thin crust, and the meteorites simply plunged into the planet, splashing up large quantities of partially solidified and molten rock. The scars in the crust quickly healed over as fresh magma bled through the surface. This created massive floods of basalt similar to those that formed the maria or lava plains on the moon. When the Earth acquired an atmosphere and ocean, intense weather systems eroded all remaining craters, and no telltale signs of the great bombardment can be found today.

The Inner Earth

As the Earth continued to cool and currents in the interior began to grow sluggish, the planet differentiated into a core, mantle and crust. The core segregated from the mantle within the first 100 million years. It is a little over half the diameter of the Earth and constitutes about one sixth of its volume and about one third of its mass. The inner core, composed of iron-nickel silicates, solidified into a sphere about 1,500 miles in diameter. The outer core is about 1,400 miles thick and is composed mostly of molten iron that flows as easily as water. The temperature of the core ranges from about 4,500 degrees Celsius on the surface to about 7,000 degrees at the boundary between the inner and outer cores, with little change in temperature toward the center.

This two-part structure is responsible for generating a strong magnetic field due to differences in rotation rate, temperature, density and chemistry between the inner and outer cores. The core and mantle are not tightly coupled, so rotating formations of fluid in the core drift westward as the Earth rotates. Since the outer core is a good electrical conductor, electric currents set up a magnetic field, which is reinforced by the rotation due to the dynamo effect, just like a giant generator.

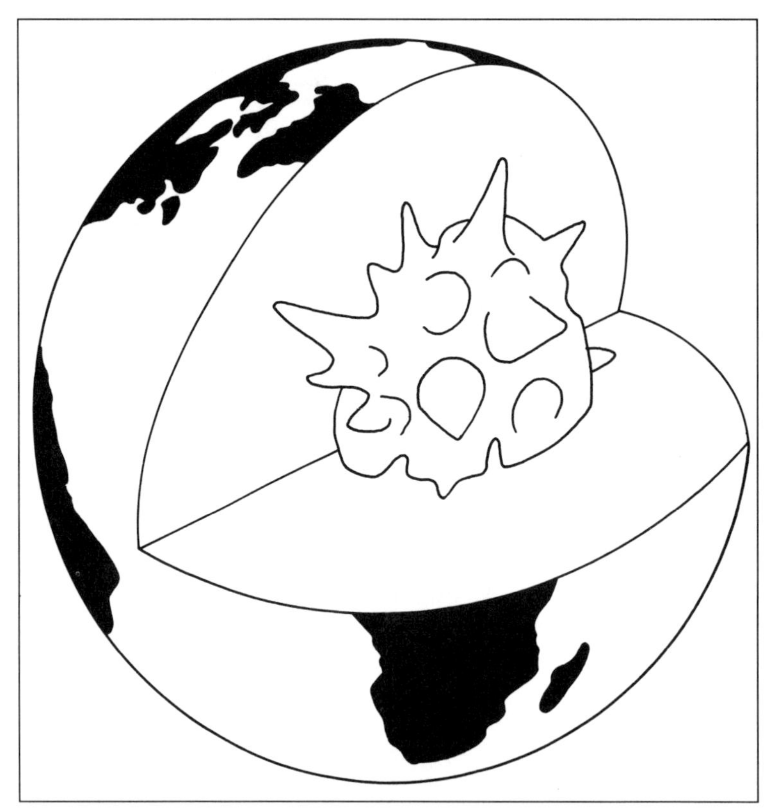

Figure 6 The core has bumps and grooves created by convection currents in the overlying mantle. Topography shown is highly exaggerated.

The surface of the fluid core is not smooth but has a broken topography (Figure 6) consisting of rises taller than Mount Everest and depressions deeper than the Grand Canyon. This relief is caused by the rising and sinking of the overlying mantle by convection currents, which apply and release pressure at certain points along the surface of the core. These mantle convection currents have lifetimes of about 100 million years and are ultimately responsible for the motion of crustal plates on the Earth's surface.

Discarded materials might be building up a slag heap on top of the core, creating "continents" at the core-mantle boundary that might interfere with the geomagnetic field as well as heat flowing from the core to the mantle. This debris might also account for a relief on the surface of the core of several hundred feet. Moreover, these continent-like masses might have roots that extend into the core, just as continents on the Earth's surface have roots that extend deep into the mantle. The shielding of heat from the core might also have a major influence on volcanism, especially concerning its role in global tectonics.

The Earth is mostly mantle, which comprises nearly half of its radius, four fifths of its volume and two thirds of its mass. The lower mantle begins at the top of the core and extends to about 400 miles beneath the Earth's surface. It is composed of primitive rock that has not changed significantly since the beginning, whereas the rocks of the upper mantle have been continuously changing composition and crystal structure with time. Mantle rocks are composed of iron-magnesium silicates in a partially molten or plastic state, which allows them to flow at a rate of perhaps several inches per year. Large-scale convection currents transport heat away from the core and distribute it along the surface of the mantle, and this transfer of heat is responsible for the operation of plate tectonics.

All activity taking place on the surface of the Earth is an outward expression of the great heat engine that drives the mantle. Continents separate, move about and collide; mountains rise skyward and

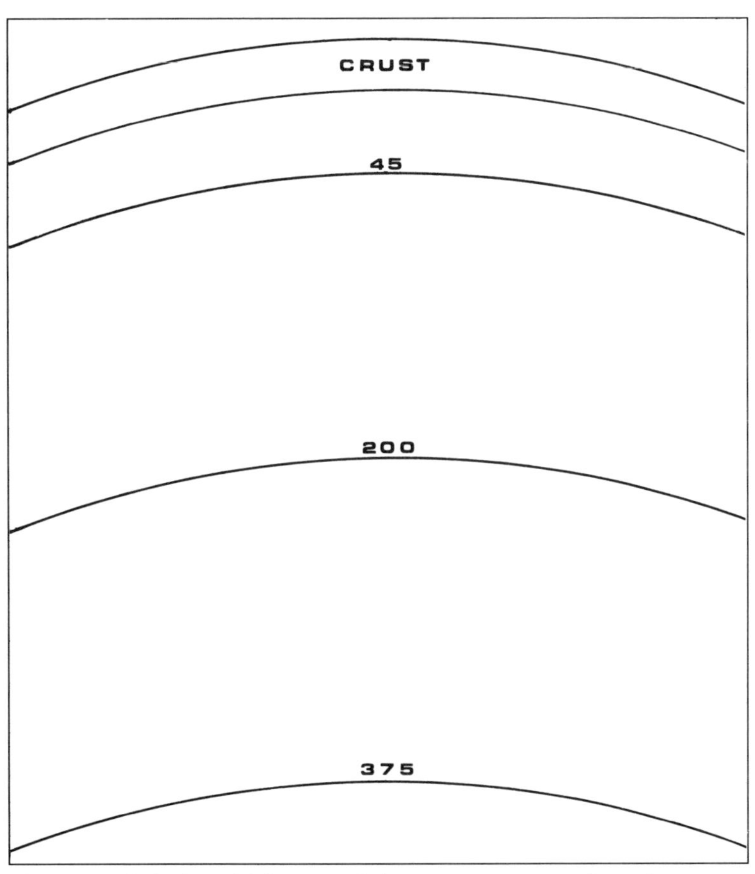

Figure 7 Relative thickness of the crust compared to the upper mantle.

ocean trenches sink downward; and volcanoes erupt and faults quake. The mantle played a major role in shaping the Earth and giving it a unique character. Gases and water vapor sweated out of the mantle by intense volcanism provided the Earth with an atmosphere and ocean. The mantle also supplied the rocks for the crust and the carbon for life.

CONTINENTAL CRUST

The Earth has the thinnest crust of all the terrestrial planets (Figure 7); even the moon's crust is thicker. The crust represents less than 1 percent of the

TABLE 1 CLASSIFICATION OF THE EARTH'S CRUST

Environment	Crust Type	Tectonic Character	Crustal Thickness (miles)	Geologic Features
Continental crust overlying stable mantle	Shield	Very stable	22	Little or no sediment, exposed Precambrian rocks
	Midcontinent	Stable	24	
	Basin-Range	Very unstable	20	Recent normal faulting, volcanism and intrusion; high mean elevation
Continental crust overlying unstable mantle	Alpine	Very unstable	34	Rapid recent uplift, relatively recent intrusion; high mean elevation
	Island arc	Very unstable	20	High volcanism, intense folding and faulting
Oceanic crust overlying stable mantle	Ocean basin	Very stable	7	Very thin sediments overlying basalts, no thick Paleozoic sediments

Environment	Crust Type	Tectonic Character	Crustal Thickness (miles)	Geologic Features
Oceanic crust overlaying unstable mantle	Ocean ridge	Unstable	6	Active basaltic volcanism, little or no sediment

Earth's radius and 0.3 percent of its mass. Including continental margins and small shallow regions in the ocean, continental crust covers about 45 percent of the Earth's surface (Tables 1 and 2). The crust is built like a layer cake with sedimentary rocks on top, granitic and metamorphic rocks in the middle and basaltic rocks on the bottom. This gives the crust a structure somewhat like a jelly sandwich with a pliable middle layer sandwiched between a hard upper crust and a hard lithosphere.

TABLE 2 COMPOSITION OF THE EARTH'S CRUST

Crust Type	Shell	Average Thickness (miles)	Percent Composition of Oxides						
			Silica	Alum	Iron	Magn	Calc	Sodi	Potas
Continental	Sedimentary	2.1	50	13	6	3	12	2	2
	Granitic	12.5	64	15	5	2	4	3	3
	Basaltic	12.5	58	16	8	4	6	3	3
Total		27.1							
Subcontinental	Sedimentary	1.8							
	Granitic	5.6							
	Basaltic	7.3	Same as above						
Total		14.7							
Oceanic	Sedimentary	0.3	41	11	6	3	17	1	2
	Volcanic Sedimentary	0.7	46	14	7	5	14	2	1
	Basaltic	3.5	50	17	8	7	12	3	<1
Total		4.5							
Average		15.4	52	14	7	4	11	2	2

The Earth did not develop a permanent crust until around 4 billion years ago, after the worst of the great meteorite bombardment. For this reason, the oldest rocks on the Earth's surface are not nearly as old as the planet itself. During the first half billion years or so, the Earth was in a fiery turmoil, and any rocks solidifying during this period soon remelted. The massive meteorite bombardment also melted the crust by impact friction. Thousands of 50-mile-wide meteorites pounded the Earth and converted 30 to 50 percent of the crust into impact basins, which later filled with water when the first rains came.

Around remote lakes and tundra of northwest Canada are rocks dating about 4 billion years old. These rocks are composed of granite, which indicates that the Earth was forming continental crust by this time. However, only 5 to 8 percent of the present continental crust was in existence between 4 and 3 billion years ago. Rocks found in southwest Greenland dating around 3.8 billion years old are composed of metamorphosed sediments that were originally laid down in a marine environment, signifying that the Earth had a significant ocean by this time. Slightly older rocks have been found in Antarctica and Africa, but for the most part few rocks on Earth date beyond 3.7 billion years old. This suggests that there were no major continents by this time but only thin slices of crust that wandered across the face of the planet driven by rapid convective motions in the mantle.

As convection curents began to slow down due to the loss of internal heat, lighter rock materials migrated toward the surface to form a basaltic scum. In essence, the crust is composed of waste products of the mantle. The reworking of this primitive crust as it was subducted into the mantle and remelted formed the first granitic rocks. The bulk of the crust is composed of oxygen, silica and aluminum, which form the granitic and metamorphic rocks that comprise the cores of the continents.

About 4 billion years ago, a giant meteorite impact (Figure 8) might have triggered the evolution of ancient continental shields, upon which the continents grew. The shields are extensive uplifted areas that are essentially bare of recent sedimentary deposits. Surrounding the shields are the platforms, which are

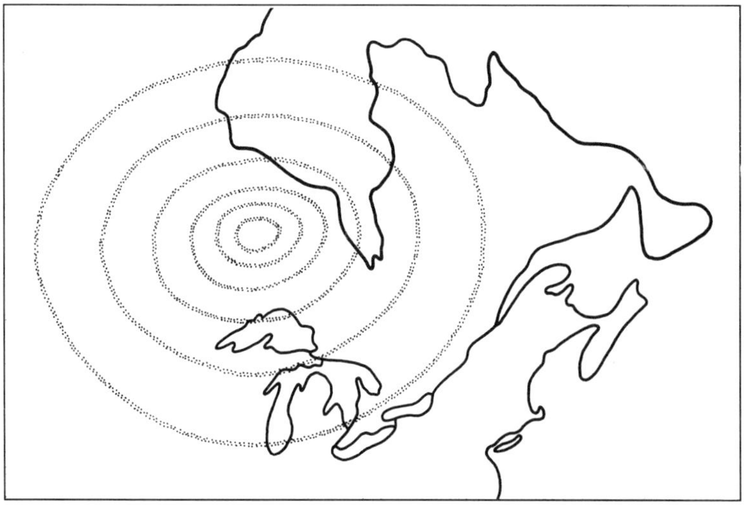

Figure 8 Location of Archean impact structure in central Ontario, Canada, thought to have triggered the formation of continents.

broad, shallow depressions of basement rock covered by nearly flat-lying sedimentary rocks. Together the shields and platforms comprise the cratons, which were the first pieces of land to appear and are found in the interiors of all continents. They are composed of ancient igneous and metamorphic rocks that are remarkably similar in composition to more recent rocks, indicating that a rock cycle was fully in place by this time.

Sediments thrust deep into the mantle were subjected to intense internal heat. The rocks either changed crystal structure or melted entirely and became magma. The buoyant magma rose to the surface in blobs called diapirs from the Greek word meaning "to pierce." If the magma broke through the surface, it gave rise to volcanic eruptions. Otherwise it remained buried in the crust, forming large granitic bodies called plutons.

The original continents were free-wheeling slivers of crust that were constantly collidiing with each other. As the Earth continued to cool, their erratic wanderings slowed down and they began to stick together, forming over a dozen protocontinents. However, their combined area was only about a tenth of today's total landmass. Eventually, all the protocontinents combined into a supercontinent.

Volcanic activity, magmatic intrusions, and rifting and patching of the continental masses built up the interior of the continent, while erosion and sedimentation built the continental margins outward. Most of the continental crust was created when two lithospheric plates collided. The collisions also greatly deformed the crust over a broad area. Worldwide continental forming events took place roughly 2.9 to 2.6 billion years ago, 1.9 to 1.7 billion years ago, 1.1 to 0.9 billion years ago and during the past 600 million years.

OCEANIC CRUST

At first, there were no landmasses to mar the Earth's watery face, and practically the entire planet was covered by a global sea up to 2 miles deep. When the first rains came, acidic rainwater reacted with metallic minerals in the crust to produce metallic salts. These were carried in solution to the sea by rivers, and seawater acquired a considerable saltiness along with other chemical substances soon after the ocean formed. This allowed the ocean to achieve chemical equilibrium early in its history.

Seawater also seeped into the fractured oceanic crust and was heated near magma chambers, then leached water-soluble compounds from the rock, rose to the ocean floor and was expelled by undersea hydrothermal vents. This exchange between the crust and the mantle continuously resupplied the ocean with the chemical substances the Earth needed to sustain life.

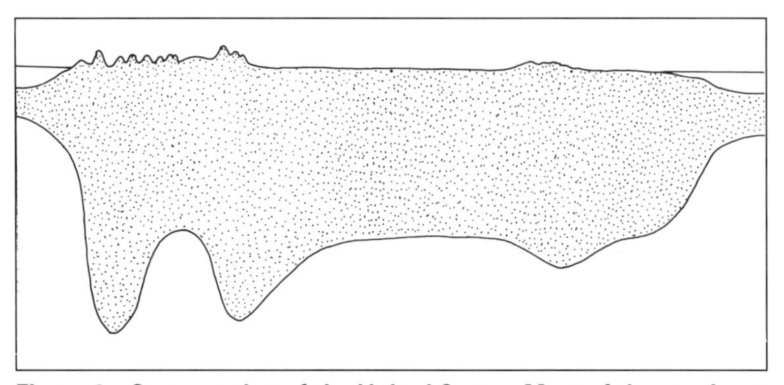

Figure 9 Cross section of the United States. Most of the continent lies below the surface.

The original crust was composed of basalt lava flows that erupted on the surface long before the ocean basins began to fill with water. Embedded in the thin basaltic crust were granitic blocks that assembled into microcontinents, or cratons. The cratons were lighter than the basalt, which allowed them to remain on the surface, drifting freely in the convection currents of the mantle as it pushed or pulled them along.

The oceanic crust is considerably thinner than continental crust, and in most places is only 3 to 5 miles thick. The oceanic crust is remarkable for its consistent thickness and temperature, averaging about 4 miles thick and varying no more than 20 degrees Celsius over most of the globe. By comparison, the continental crust averages 25 to 30 miles thick, and in the vicinity of mountain ranges it reaches depths of 45 miles. Continental crust has an average density of 2.7 times the density of water, whereas oceanic crust is 3.0 times as dense. Because the mantle's density is 3.4, the continental and oceanic crusts remain afloat above the mantle.

Like icebergs, most of the crust resides beneath the surface (Figure 9). The long-lived continental roots that underlie mountain ranges can extend downward as much as 250 miles into the upper mantle. Continental crust is also 20 times older than oceanic crust, which is no older than 180 million years, because oceanic crust is recycled back into the mantle, and almost all the seafloor has since disappeared into the Earth's interior to provide the raw materials for the continued growth of the continents.

New oceanic crust created at spreading ridges starts out relatively thin and eventually thickens by the underplating of new lithosphere from the upper mantle and the accumulation of overlying sediment layers. By the time the oceanic crust spreads out as wide as the Atlantic Ocean, the portion near continental margins where the ocean is the deepest is over 50 miles thick. Eventually, the oceanic crust becomes too thick and heavy to remain on the surface and subducts into the mantle, where it melts to provide material for new crust.

When a continental plate collides with an oceanic plate, the latter bends downward and subducts into the mantle. The oceanic plate remelts in the Earth's interior, acquires new minerals from the mantle and reemerges at volcanic spreading centers, most of which are in the ocean (Figures 10A and 10B). This rejuvenates the ocean crust while helping to spread the continents farther apart. Sediments deposited on the ocean floor along with water trapped between sediment grains are also caught in the subduction zones. But because of their lower melting points and lesser density, they rise to the surface to supply volcanoes with magma and recycled seawater.

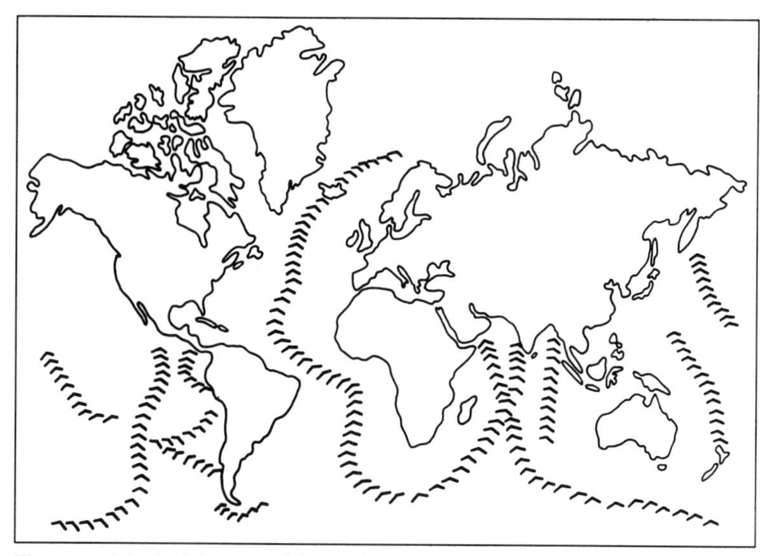

Figure 10A Midocean ridges that wind around the world's ocean basins are composed of individual volcanic spreading centers.

Rocks in the asthenosphere, which is the fluid portion of the upper mantle, heat up and become plastic, enabling them to slowly rise by convection. After millions of years they reach the topmost layer of the mantle, or lithosphere. When the rocks reach the underside of the lithosphere, they spread out laterally, cool and descend back into the deep interior of the Earth. The constant pressure against the bottom of the lithosphere creates fractures that weaken it. As the convection currents flow out on either side of the fracture, they carry the separated halves of the lithosphere along with them, and the rift continues to widen.

As the pressure decreases, the rocks melt and rise through the fracture. The molten magma passes through the lithosphere until it reaches the oceanic crust, where it forms magma chambers that further press laterally against the oceanic crust, and the rift widens. The magma chambers also provide molten lava that pours out of the trough between the two ridge crests, adding new material to both sides of a spreading ridge. This generates about 3 cubic miles of new oceanic crust each year. The pressure of the upwelling magma forces the ridge farther apart, pushing the ocean floor and the lithosphere upon which it rides away from the spreading ridge.

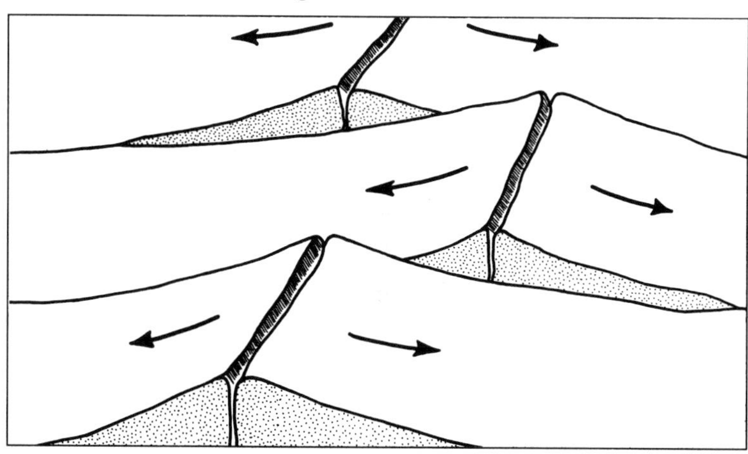

Figure 10B Spreading centers spread segments of the seafloor apart.

The mantle below the spreading centers, where new oceanic crust is created, consists mostly of peridotite, a strong, dense rock composed of iron and magnesium silicates. As the peridotite melts on its 30- to 40-mile journey to the base of the crust, a portion of it becomes highly fluid basalt. Basalt is the most common magma erupted on the Earth's surface, and about 5 cubic miles of basaltic magma is removed from the mantle and added to the crust every year. Most of this volcanism occurs on the ocean floor at spreading centers, where the oceanic crust is being pulled apart. Gabbros, which have higher amounts of silica, solidify out of the basaltic melt and accumulate in the lower layer of the crust.

CRUSTAL PLATES

The Earth's outer shell is fashioned out of seven major and several minor tectonic plates (Figure 11). The plates are composed of the upper brittle crust and the upper brittle mantle called the lithosphere, which averages about 60 miles thick. The lithosphere consists of the rigid outer layer of the mantle

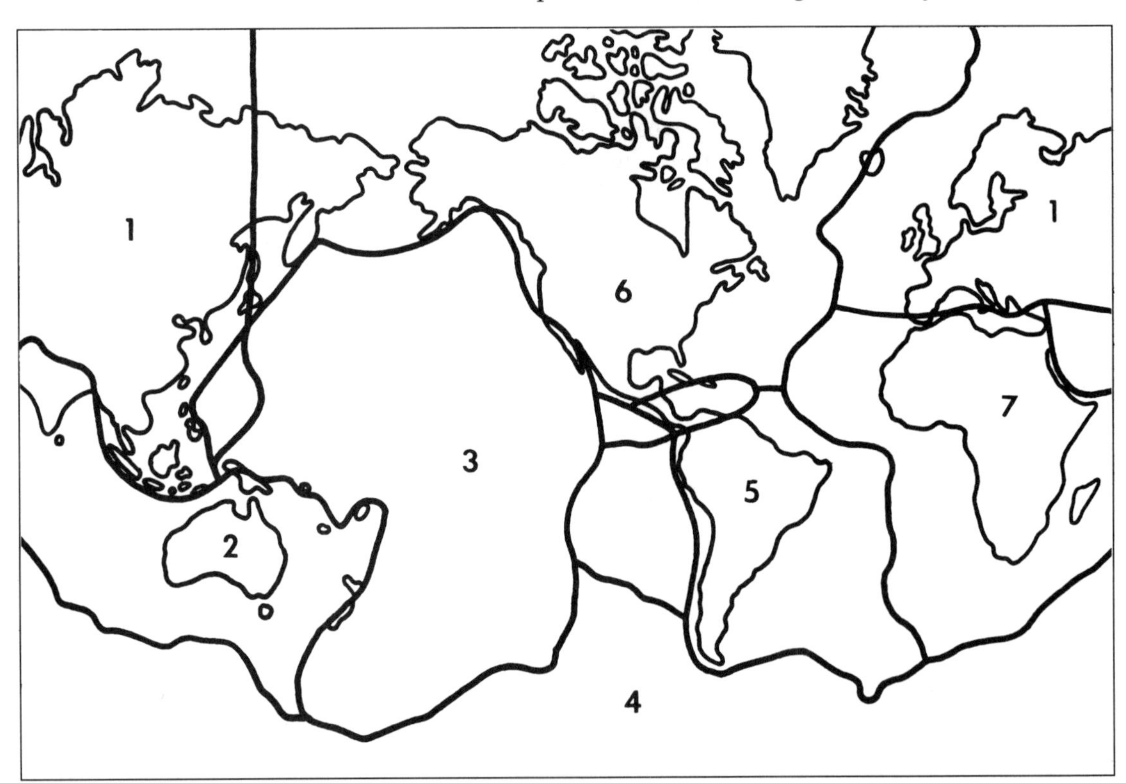

Figure 11 The lithospheric plates that comprise the Earth's crust: 1. Eurasian plate, 2. Australindian plate, 3. Pacific plate, 4. Antarctic plate, 5. South American plate, 6. North American plate, 7. African plate.

and the overlying continental or oceanic crust and rides on the semimolten layer of the upper mantle called the asthenosphere. This structure is important for the operation of plate tectonics, without which the crust would become jumbled up slabs of rock like arctic pack ice.

The lithospheric plates act like rafts riding on a sea of molten rock and carry the crust around the surface of the globe. The plates diverge at spreading ridges and converge at subduction zones, which are depicted on the ocean floor as deep-sea trenches, where the plates dive into the mantle and remelt. The plates and oceanic crust are continuously recycled through the mantle, but the continental crust due to its geater buoyancy remains on the surface.

Figure 12 The rifting of the Red Sea, viewed from the Gemini spacecraft. Courtesy of NASA

Rifting and patching of the continents has taken place for the past 2.7 billion years and possibly much longer than that. The best evidence for the rifting of continents is found in the East African rift system, which has not yet fully ruptured. When it does, however, the present continental rift will be replaced by an oceanic rift. This type of rifting is presently taking place in the Red Sea, where Africa and Saudi Arabia are pulling away from each other (Figure 12).

The transition from a continental rift to an oceanic rift is accompanied by block faulting, whereby blocks of continental crust drop along extensional faults, where the crust is being pulled apart, resulting in a deep rift valley and a thinning of the crust. Upwelling of molten rock from the mantle further weakens the crust, and convection currents pull apart the divided segments of crust. Therefore, the engine that drives the birth and development of rifts and, consequently, the breakup of continents and the formation of ocean basins ultimately comes from the Earth's interior.

2

BASEMENT ROCK

Underlying all other rocks on the Earth's surface is a thick layer of basement complex, composed of ancient granitic and metamorphic rocks that have been in existence for nine-tenths of the Earth's history. These rocks form the nuclei of the continents and first appeared during a period of mantle segregation and outgassing, which created the crust along with the atmosphere and ocean. One remarkable feature about these rocks is that, despite their great age, they are similar to more recent rocks, signifying that geologic processes began quite early and had a long and productive life.

PRECAMBRIAN SHIELDS

The first 4 billion years on Earth, or about 90 percent of geologic time, is called the Precambrian, the longest and least understood period of the Earth's history. The Precambrian is divided into the Archean eon, from 4.6 to 2.5 billion years ago, and the Proterozoic eon, from 2.5 to 0.6 billion years ago. The boundary between the Archean and Proterozoic reflects major differences in the character of rocks older than 2.5 billion years as compared with later rocks.

Figure 13 Lava pool and fountains on Kilauea, Hawaii, in July 1961. Photo by D. H. Richter, courtesy of USGS

The Archean was a time when the Earth's interior was hotter, the crust was thinner and more unstable, and tectonic plates were more mobile. The Earth was subjected to extensive volcanism and intense meteorite bombardments. The first 700 million years of Archean time is missing from the geologic record because no rocks could have survived the tumultuous activity taking place on the Earth's surface during this time.

Heavy turbulence in the mantle with a heat flow three times greater than it is today produced violent agitation on the surface, resulting in a sea of molten and semimolten rock broken up by giant fissures, from which fountains of lava spewed skyward (Figure 13). Jangled pieces of solid crust jostled against each other, broke up and slid under the magma. Meteorites up to 60 miles wide hammered the infant planet and were quickly swallowed up by the boiling cauldron as they plunged into the Earth.

When the Earth was about 500 million years old, most of the short-lived, energetic radioactive elements in the mantle decayed into stable daughter products as the mantle gradually began to cool. This resulted in the creation of a permanent crust composed of a thin layer of basalt embedded with scattered blocks of granite called "rockbergs." Slices of granitic crust combined into stable bodies of basement rock, upon which all other rocks were deposited.

The basement rocks formed the cores of the continents and are presently exposed in broad, low-lying, domelike structures called shields (Figure 14). The shields are extensive uplifted areas surrounded by sediment-covered bedrock called continental platforms, which are broad, shallow depressions of basement complex filled with nearly flat-lying sedimentary rocks. Many shields, such as the Canadian Shield, which covers most of eastern Canada, are fully exposed in areas that were ground down by flowing ice sheets during the last ice age. The exposure of the Canadian Shield from Manitoba to Ontario is attributed to uplifting of the crust by a mantle plume and the erosion of sediments in the uplifted area.

Archean greenstone belts are dispersed among and around the shields. They consist of a jumble of metamorphosed (recrystallized) lava flows and sediments possibly derived from island arcs (chains of volcanic islands on the edges of subduction zones) that were caught between colliding continents. Their green color is derived from the mineral chlorite, a greenish form of mica. The existence of greenstone belts is evidence that plate tectonics might have operated as early as the Archean.

Ophiolites, named from the Greek *ophis*, meaning "serpent," due to their mottled green color, date as far back as 3.6 billion years. These rocks were also caught in the greenstone belts, as slices of ocean floor were shoved up on the continents by drifting plates. Therefore, ophiolites are among the best evidence for ancient plate motions. They are vertical cross sections of oceanic crust that were peeled off during plate collisions and plastered onto the continents. This gave rise to a linear formation of greenish volcanic rocks along with light-colored masses of granite and gneiss, which are common igneous and metamorphic

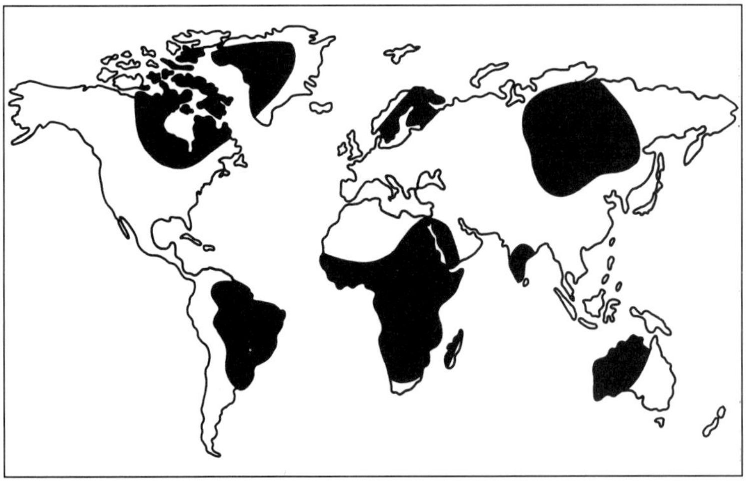

Figure 14 Location of continental shields, the foundations upon which the continents grew.

rocks respectively. Pillow lavas, which are tubular bodies of basalt extruded undersea, are also found in the greenstone belts, signifying that the volcanic eruptions took place on the ocean floor. Many ophiolites contain ore-bearing rocks, which are important mineral resources throughout the world.

Greenstone belts are found in all parts of the world and occupy the ancient cores of the continents (Figure 15). They span across an area of several hundred square miles and are surrounded by immense expanses of gneisses, which are the metamorphic equivalents of igneous rocks and the predominant Archean rock types. The best known of the greenstone belts is the Swaziland sequence in the Barberton Mountain Land of southeastern Africa. It is over 3 billion years old and reaches a thickness of nearly 12 miles.

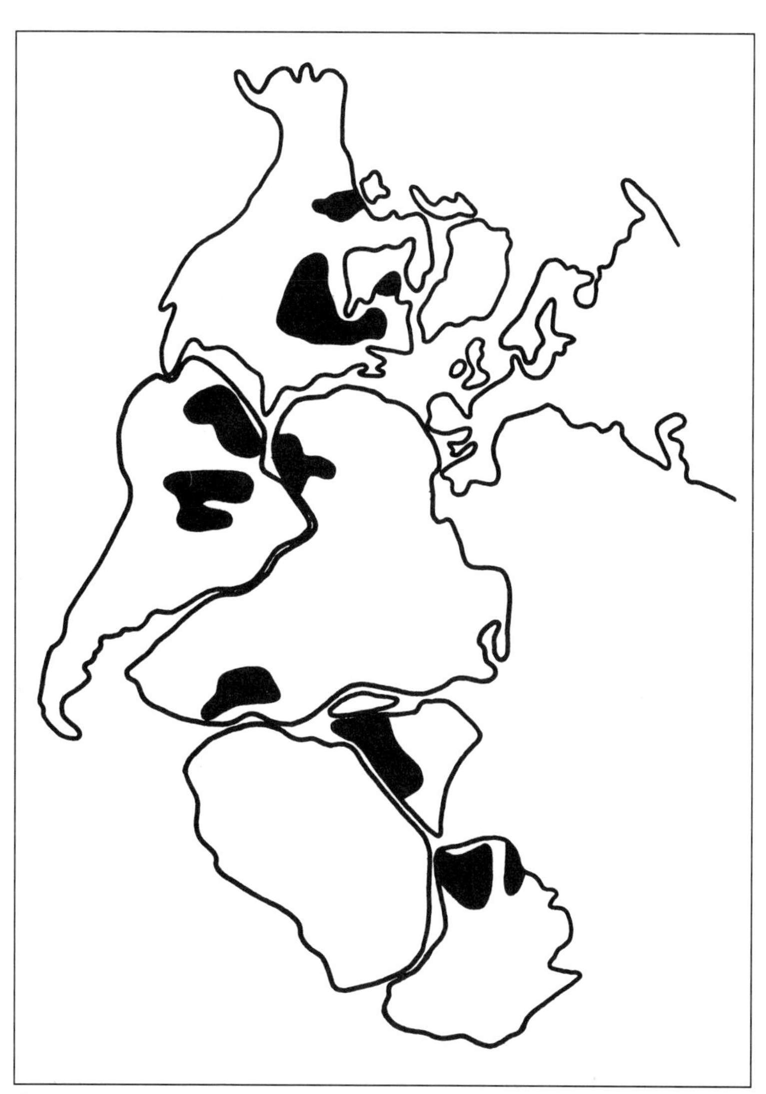

Figure 15 Archean greenstone belts occupy the ancient cores of the continents.

Greenstone belts are of particular interest to geologists, not only as evidence for Archean plate tectonics but also because they hold the majority of the world's gold deposits. Most of the South African gold mines are in greenstone belts, and the Kolar belt in India holds the richest gold deposits in the world. Since greenstone belts are essentially Archean in age, their disappearance around 2.5 billion years ago marks the end of the Archean eon.

The abundance of chert, a dense, extremely hard sedimentary rock, in deposits older than 2.5 billion years indicates that most of the crust was deeply submerged during this time. Cherts are composed of microscopic grains of silica. Most Precambrian cherts are thought to be chemical sediments precipitated from silica-rich water in deep oceans. Ancient cherts dating 3.5 billion years old also contain microfilaments, which are believed to have a bacterial origin and therefore provide some of the oldest evidence of life.

The 3.8 billion-year-old metamorphosed marine sediments of the Isua Formation in a remote mountainous area in southwest Greenland (Figure 16) provide evidence that oceans were in existence by this time. The seas contained abundant dissolved silica, which leached out of volcanic rock pouring onto the ocean floor. Modern ocean water is deficient in silica because organisms like sponges and diatoms extract it to build their skeletons. When the organisms die, their skeletons build up massive deposits of diatomaceous earth.

The Proterozoic eon marked a major change from the turbulent Archean. At the beginning of the Proterozoic, nearly three quarters of the present landmass was in existence. The continental crust had an average thickness of between 15 and 25 miles, near what it is today. Continents were more stable and combined into a supercontinent composed of Archean cratons, which are pieces of granitic crust that formed the cores of the continents. Many of the Archean cratons throughout the world were assembled around the same time. Some of the original cratons formed within the first 1.5 billion years of the Earth's existence but totaled only about one tenth of the present landmass.

Figure 16 Location of the Isua Formation in southeastern Greenland.

Seven major cratons came together to form the proto–North American continent called Laurentia

(Figure 17), most of which was assembled some 2 billion years ago, making it one of the oldest continents on Earth. At Cape Smith on the Hudson Bay lies a 2-billion-year-old piece of oceanic crust that was squeezed onto the land, a telltale sign that continents collided and closed an ancient ocean. Arcs of volcanic rock also weave through central and eastern Canada down into the Dakotas. In a region between Canada's Great Bear Lake and the Beaufort Sea lies the roots of an ancient mountain range that run through the basement rock. The mountains were formed by the collision of North America and an unknown landmass between 1.2 and 0.9 billion years ago.

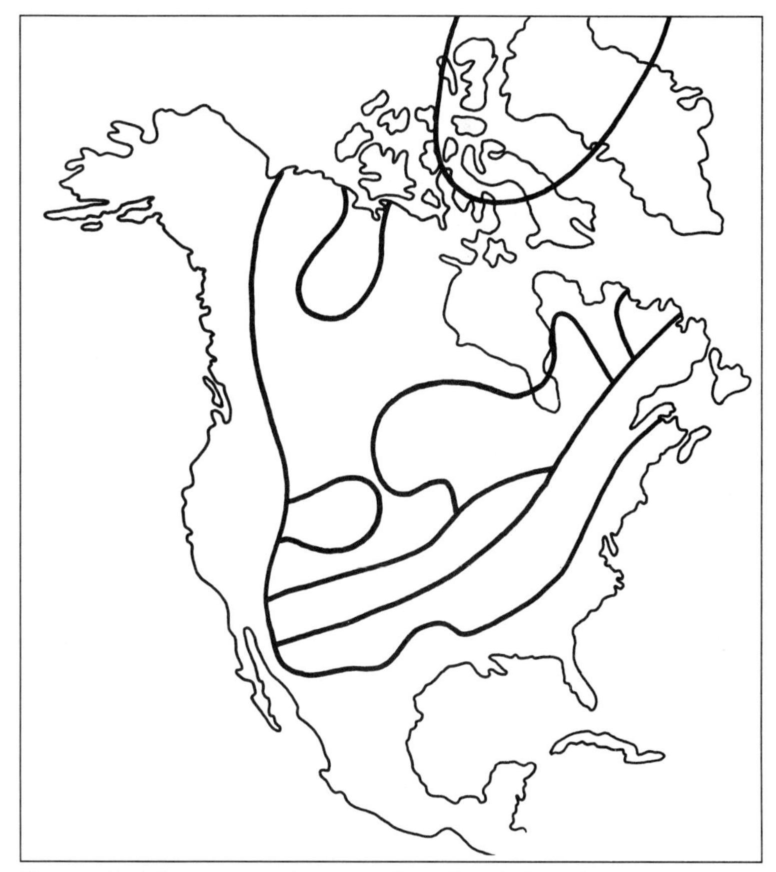

Figure 17 The cratons that constitute North America.

Continental collisions continued to add a large area of new crust to the growing proto–North American continent. The better part of the continental crust underlying the United States from Arizona to the Great Lakes to Alabama formed in one great surge of crustal generation between 1.9 and 1.7 billion years ago that has since been unequaled in North America. The assembled North American continent was stable enough to resist a billion years of jostling and rifting and continued to grow as bits and pieces of continents and island arcs were plastered to its sides. The Superior Province just north of the Great Lakes consists of island arcs and sediment sandwiched together and broken around the edges.

This was possibly the most energetic period of tectonic activity and crustal generation in the Earth's history. The presence of volcanic rock near the eastern edge of North America is a sign that a giant rift ripped through the continent, implying it was part of a supercontinent in the early Proterozoic. Toward the end of the Proterozoic around 600 million years ago, another supercontinent with North America at its core broke into possibly four or five major continents, although their geographies were not the same as they are today. About 1.1 billion years ago, a great volcano-filled rift valley split the continent from what is now Kansas to Lake Superior, possibly as a prelude to continental breakup.

By the beginning of the Proterozoic, most of the material presently locked in sedimentary rocks was already at or near the surface, with ample sources of Archean rocks for erosion and redeposition. Most Proterozoic sediments are composed of sandstones and siltstones derived from Archean greenstones. Conglomerates, which are consolidated equivalents of sand and gravel, were particularly abundant in the Proterozoic. The Proterozoic is also known for its widespread terrestrial redbeds, so named because their sediment grains were cemented by red iron oxide. Their appearance about 1 billion years ago signifies that the atmosphere and ocean contained significant amounts of oxygen by this time.

The weathering of primary rocks also yielded solutions of calcium carbonate, magnesium carbonate, calcium sulfate and sodium chloride, which precipitated consecutively into limestone, dolomite, gypsum and

Figure 18 The Grand Canyon showing the Precambrian Vishnu Schist, overlain by younger rocks of the Grand Canyon Series and Tonto Group, Coconino County, Arizona. Photo by E. D. McKee, courtesy of USGS

halite. The Mackenzie Mountains of northwest Canada contain deposits of dolomite 6,500 feet thick. These were mainly chemical precipitates rather than the deposits of biological activity because shell-producing organisms had not yet evolved. Carbonate rocks such as limestone and chalk that were formed chiefly by organic process involving shells and skeletons of simple organisms became much more common during the late Proterozoic, beginning about 700 million years ago. Before this, however, these rocks were relatively rare due to the scarcity of lime-secreting organisms.

CRATONS

Some of the oldest known rocks of North America are the 2.5-billion-year-old granites of the Canadian Shield, which covers eastern Canada and extends down into Wisconsin and Minnesota. Some of the best exposures

Figure 19 The Rocky Mountains represent upraised portions of the Earth's interior. Photo by George A. Grant, courtesy of National Park Service

of ancient rocks in the United States are the 1.8-billion-year-old metamorphic rocks on the bottom of the Grand Canyon (Figure 18). In northern Arizona, over a mile of sedimentary rocks overlie the bedrock of the Grand Canyon. The oldest of these rocks is about 800 million years old, leaving a billion years of geologic history unaccounted for. During this time, the floor of the Grand Canyon was worn by erosion, creating a gap in recorded time known as a hiatus.

On top of the Grand Canyon floor, marine sediments slowly were laid. The continuous buildup of sediments caused the ancient seafloor to subside due to the increased weight. In a fraction of the time it took to deposit the sediments, a gradual upheaval brought them to their present elevation, during which time the swift-flowing Colorado River gouged out layer upon layer of rock, exposing the raw earth below.

The cores of the world's mountain ranges also contain some of the oldest rocks, which were once buried deep in the crust but have since been raised to the surface (Figure 19). As the mountains were pushed upward, huge blocks of granite were thrust up by tectonic forces originating deep within the Earth.

The continents are composed of odds and ends of ancient cratons, which were the first pieces of crust to appear and are found in the heart of every continent. The cratons were slivers of crust that collided with and bounced off each other. As the Earth aged and continued to cool, the cratons slowed their erratic wanderings and began to stick to one another, forming over a dozen protocontinents. They are over 2.5 billion year old and range in size from smaller than the state of Texas to about one fifth of the area of present-day North America. As a whole, however, they comprised only about a tenth of today's total landmass.

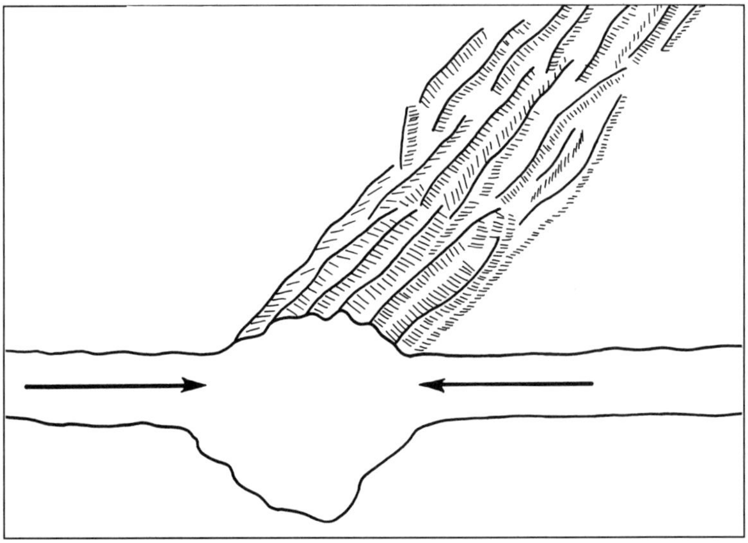

Figure 20 Mountain ranges are formed by the collision of two continental plates.

When the continents collided, they crumpled the crust, forcing up mountain ranges at the point of contact (Figure 20). The sutures joining the landmasses are still visible as cores of ancient mountains that are over 2 billion years old. Caught between the converging cratons was an assortment of debris swept up by the drifting continents, including sediments from continental shelves and the ocean floor, stringers of volcanic rock and small scraps of continents, all sliced up by faults. In addition, pieces of ocean crust called ophiolites were thrust up on land along with blueschists (Figure 21), which are metamorphosed rocks of subducted ocean crust shoved up on the continents.

Figure 21 An outcrop of retrograde blueschist rocks in the Seward Peninsula region, Alaska. Photo by C. L. Sainsbury, courtesy of USGS

The cratons contain some of the world's oldest rocks and date as far back as 4 billion years. They are composed of highly altered granite and metamorphosed marine sediments, and lava flows. The rocks originated from intrusions of magma into the primitive ocean crust. This allowed the magma to cool slowly and separate into a light component, which rose toward the surface, and a heavy component, which settled to the bottom of the magma chamber. Some magma also seeped through the crust, where it poured out as lava on the ocean floor. Successive intrusions and extrusions of magma built up the crust until it finally broke the surface of a global sea.

The cratons were highly mobile and moved about freely on the molten rocks of the asthenosphere, the fluid portion of the upper mantle. The independent, freewheeling minicontinents periodically collided with each other. The collisions caused a slight crumpling at the leading edges of the cratons, forming small parallel mountain ranges that were perhaps only a few hundred feet high.

Volcanoes were highly active on the cratons, and lava and ash continued to build the landmasses upward and outward. New crustal material was also added to the interior of the cratons by magmatic intrusions composed of molten crustal rocks that were recycled through the upper mantle. This cooled the mantle, slowing down the cratons. As the cratons grew more sluggish, they developed a greater tendency to cling to each other.

All the cratons eventually coalesced into a large landmass several thousand miles wide. The average rate of continental growth since the Earth began was perhaps as much as 1 cubic mile a year. The constant rifting and patching of the interior along with sediments deposited along the continental margins continued to build the supercontinent outward until eventually its area was nearly equal to the total area of all modern continents.

TERRANES

The cratons are patchwork mixtures, consisting of crustal pieces known as terranes, which were assembled into geologic collages. The term *terrane* should not be confused with *terrain,* which means landform. Terranes are usually bounded by faults and are distinct from their geologic surroundings. The boundary between two or more terranes is called a suture zone. The composition of terranes generally resembles that of an oceanic island or plateau, and others are composed of a consolidated conglomerate of pebbles, sand and silt that accumulated in an ocean basin between colliding crustal fragments.

Figure 22 Radiolarian skeletons can determine the histories of terranes by the various shapes of these single-celled organisms.

Terranes exist in a variety of shapes and sizes from small slivers to subcontinents such as India, which is a single great terrane. Most terranes are elongated bodies that tend to deform when they collide and accrete to a continent. The assemblage of terranes in China is being stretched and displaced in an east-west direction due to the continuing squeeze India is exerting on southern Asia as it raises the Himalayas. North of the Himalayas lies a belt of ophiolites, which appears to mark the boundary between the sutured continents.

The terranes range in age from less than 200 million years old to well over 1 billion years old. Their ages are determined by the study of fossil radiolarians (Figure 22), which are marine protozoans that built skeletons of silica and were abundant from about 500 to 160 million years ago. Different species also defined specific regions of the ocean where the terranes originated.

Suspect terranes, so-named because of their exotic origins, are fault-bounded blocks with geologic histories that are totally different from those of neighboring terranes and of adjoining continental masses. Terrane boundaries are commonly marked by ophiolite belts, consisting of marine sedimentary rocks, pillow basalts, sheeted dike complexes, gabbros and periodotites. Suspect terranes were displaced over great distance before finally being accreted to a continental margin. Some North American suspect terranes have a western Pacific origin and were displaced thousands of miles to the east.

Until about 250 million years ago, the western edge of North America ended near present-day Salt Lake City. Over the last 200 million years, North America has expanded by more than 25 percent during a major pulse of crustal growth. Much of western North America was assembled from oceanic island arcs and other crustal debris that were skimmed off the Pacific plate as the North American plate continued heading westward. The accreted terranes played a major role in the creation of mountain chains along convergent continental margins. For example, the Andes might have been thrown up by the accretion of oceanic plateaus along the continental margin of South America.

Northern California is a jumble of crust assembled some 200 million years ago. A nearly complete slice of ocean crust shoved up on the continents by drifting plates sits in the middle of Wyoming and is at least 2.7 billion years old. Many of the terranes in western North America have rotated in a clockwise direction as much as 70 degrees or more, with the oldest terranes having the greatest rotations.

The entire state of Alaska is an agglomeration of terranes that were pieces of an ancient ocean that preceded the Pacific, called the Panthalassa. The entire state is an assemblage of some 50 terranes that were set adrift over the past 160 million years by the wanderings and collisions of crustal plates, pieces of which are still arriving from the south. California west of the San Andreas Fault has been drifting northward for millions of years, and in another 50 million years it will wander as far north as Alaska.

A large portion of the Alaskan panhandle, known as the Alexander terrane, began its existence as part of eastern Australia some 500 million years ago. About 375 million years ago, it broke free from Australia, traversed the Pacific Ocean, stopped briefly at the coast of Peru, sliced past California swiping some of the Mother Lode gold belt, and bumped into the North American continent around 100 million years ago.

Figure 23 The San Francisco peninsula. The San Andreas Fault runs vertically through the center of the photograph. Photo by R. E. Wallace, courtesy of USGS

The actual distances terranes can travel varies considerably. Basaltic seamounts that accreted to the margin of Oregon moved from nearby offshore, while similar rock formations around San Francisco, California (Figure 23), came halfway across the Pacific Ocean. The city is actually built on three different and distinct rock units. At their usual rate of travel, terranes could make a complete circuit of the globe in only about 500 million years.

Terranes created on an oceanic plate retain their shape until they collide and accrete to a continent. They are then subjected to crustal movements that modify their shape. The terranes that make up the Brooks Range, the spine of northern Alaska, are great sheets stacked on top of each other (Figure 24). Along the mountain ranges in western North America the terranes are elongated bodies due to the slicing of the crust by a network

Figure 24 Steeply dipping Paleozoic rocks of the Brooks Range, Anaktuvuk district, northern Alaska.
Photo by J. C. Reed, courtesy of U.S. Navy and USGS

of northwest-trending faults such as the San Andreas Fault in California, which has undergone some 200 miles of displacement in the last 25 million years.

CRYSTALLINE ROCK

The first rocks to appear on Earth were igneous rocks, which formed directly from molten magma. Most igneous rocks arise from new material in the mantle, some are derived from the subduction of oceanic crust into the mantle and others come from the melting of continental crust. The first two types continuously build the continents, and the latter type adds nothing to the total volume of continental crust.

Igneous rocks are mostly silicates, which are compounds of silica and oxygen that contain metal ions. They are not simple chemical compounds, however, because their composition is not determined by a fixed ratio of atoms. Often two or more compounds are present in a solid solution with

Figure 25 Mount Hood Volcano in the Cascade Range, Hood River County, Oregon. Photo by M. V. Shulters, courtesy of USGS

each other. In this manner, the components can be mixed in any ratio over a wide range.

If magma extrudes onto the Earth's surface either by a fissure eruption, the most prevalent kind, or a volcanic eruption, which builds majestic mountains (Figure 25), it produces a variety of rock types depending on the source material, which in turn controls the type of eruption. Ejecta from volcanoes have a wide range of chemical, mineral and physical properties (Table 3). Nearly all volcanic products are silicate rocks, containing various amounts of other elements. Basalts are relatively low in silica and have a high content of calcium, magnesium and iron. Magmas that have larger amounts of silica, sodium and potassium along with lesser amounts of magnesium and iron form both rhyolites, which contain mostly quartz grains, and andesites, which contain mostly feldspar grains.

Igneous rocks are classified by their mineral content and texture (Table 4), which in turn are governed by the degree of separation and rate of cooling of the magma. The most common crystalline rocks are granites and

TABLE 3 CLASSIFICATION OF VOLCANIC ROCKS

Property	Basalt	Andesite	Rhyolite
Silica content	Lowest about 50%, basic rock	Intermediate about 60%	Highest more than 65%, acid rock
Dark mineral content	High	Intermediate	Low
Typical minerals	Feldspar Pyroxene Olivine Oxides	Feldspar Amphibole Pyroxene Mica	Feldspar Quartz Mica Amphibole
Density	High	Intermediate	Low
Melting point	High	Intermediate	Low
Molten rock viscosity at the surface	Low	Intermediate	High
Tendency to form lavas	High	Intermediate	Low
Tendency to form pyroclastics	Low	Intermediate	High

TABLE 4 CLASSIFICATION OF ROCKS

Group	Characteristics	Environment
Igneous		
Intrusives	Granite. Mostly quartz and potassium feldspar with mica, pyroxene and amphibole	Deep-seated coarse-grained pluton
	Syenite. Mostly potassium feldspar with mica, pyroxene and amphibole	Deep-seated medium-grained pluton
	Monzonite. Plagioclase and potassium feldspar with mica, pyroxene and amphibole	Deep-seated coarse-grained pluton
	Diorite. Mostly plagioclase and quartz with abundant mica, pyroxene and amphibole	Deep-seated coarse-grained pluton
	Gabbro. Equal amounts of plagioclase and mica, pyroxene and amphibole	Intermediate depth medium- to coarse-grained pluton
	Peridotite. Mostly olivine, pyroxene and amphibole with little plagioclase	Very deep seated medium- to fine-grained pluton
Extrusives	Rhyolite. Mostly quartz and potassium feldspar with mica, pyroxene and amphibole	Fine-grained fissure or volcanic eruption
	Andesite. Mostly plagioclase and quartz with abundant mica, pyroxene and amphibole	Fine-grained fissure or volcanic eruption
	Basalt. Equal amounts of plagioclase and mica, pyroxene and amphibole	Fine-grained fissure or volcanic eruption

Group	Characteristics	Environment
Metamorphic		
Foliated	Gneiss. Mostly quartz and feldspar with mica and amphibole	Coarse-grained, deep-seated
	Schist. Mostly mica and platy minerals with less quartz and feldspar	Coarse-grained, deep-seated
	Phyllite. Micaceous rock intermediate between schist and slate	Medium-grained, moderate depth
	Slate. Feldspar quartz and micaceous minerals	Fine-grained, moderate depth
Nonfoliated	Hornfels. Metamorphic clay material	Contact with hot magma bodies
	Marble. Metamorphic carbonates	Coarse-grained, deep-seated
	Quartzite. Metamorphic sandstone	Fine-grained, deep-seated
Sedimentary		
Clastic	Conglomerate. Fragments of rounded gravel-size sediments	River and glacial deposits
	Breccia. Fragments of angular gravel-size sediments	River and volcanic deposits
	Sandstone. Coarse-grained quartz and feldspar with minor accessory minerals	Marine and river deposits
	Siltstone. Fine-grained quartz and feldspar with minor accessory minerals	Marine, lake, and river deposits
	Shale. Very fine grained sediments, mostly feldspar	Marine and lake deposits

Group	Characteristics	Environment
Nonclastic	Limestone. Calcium carbonate often with skeletal fragments	Marine and lake deposits
	Dolomite. Calcium magnesium carbonate	Marine deposits and veins
	Gypsum. Hydrous calcium sulfate	Near-shore brine pools
	Chalcedony. Microscopic silica	Deep marine and groundwater

metamorphics, which form most of the interiors of the continents. The texture of the granitic rocks is controlled by the rate of cooling, with the slowest rates giving rise to the largest crystals and the more rapid rates providing smaller grains. Most igneous rocks are aggregates of two or more minerals. Granite, for example, is composed almost entirely of quartz and feldspar with a minor portion of other minerals. Granitic rocks form deep within the crust, and crystal growth is controlled by the magma cooling rate and the available space.

Large crystals probably form late in the crystallization of a large magma body such as a batholith. The best known of these include the Sierra Nevada, the California and the Andean batholiths. Such large bodies provide new additions to the crust. Large crystals might also form in the presence of volatiles such as water and carbon dioxide, permitting them to grow in a smaller volume. As a magma body slowly cools, possibly taking 1 million years or more, the crystals are able to grow directly out of the fluid melt or out of the volatile magmatic fluids that invade the surrounding rocks.

If a granitic rock develops extremely large crystals, it is called a pegmatite, which is the major source of single-mineral crystals throughout the world. Granite pegmatites are known for having crystals of exceptional size. They also contain rare minerals with smaller crystals. The granite associated with pegmatites often consists of quartz rods in a matrix of feldspar combined in an interlocking angular pattern called graphic granite.

Pegmatites are composed mainly of great masses of quartz and feldspar, which are the dominant minerals in granite. Many minerals that normally form only microscopic crystals in granite often form very large crystals in pegmatites. Pegmatites are found in many localities where well-exposed granite outcrops exist, but they are especially prevalent in the mountainous eastern and western regions of the United States.

Igneous and sedimentary rocks subjected to the intense temperatures and pressures of the Earth's interior, by heat generated near magma bodies, by shear pressures from Earth movements, or by strong chemical reactions

TABLE 5 CRUSTAL ABUNDANCE OF ROCK TYPES AND MINERALS

Rock Type	Percent Volume	Minerals	Percent Volume
Sandstone	1.7	Quartz	12.0
Clays and shales	4.2	Potassium feldspar	12.0
Carbonates	2.0	Plagioclase	39.0
Granites	10.4	Micas	5.0
Grandiorite Quartz diorite	11.2	Amphiboles	5.0
Syenites	0.4	Pyroxenes	11.0
Basalts Gabbros Amphibolites Granulites	42.5	Olivine	3.0
		Sheet silicates	4.6
Ultramafics	0.2	Calcite	1.5
Gneisses	21.4	Dolomite	0.5
Schists	5.1	Magnetite	1.5
Marbles	0.9	Other	4.9

that do not result in melting, produce metamorphic rocks. Metamorphism causes dramatic changes in texture, mineral composition, or both. It produces new textures by recrystallization, which causes minerals to grow into larger crystals. New minerals are also created by recombining chemical elements to form new associations. Water and gases from nearby magma bodies also aid in chemical changes that occur in rocks by conveying chemical elements from one place to another.

Heat is the primary agent for recrystallization, and often deep burial is required to generate the temperatures and pressures required for extensive metamorphism. Varying degrees of metamorphism are also achieved at shallower depths in geologically active areas with higher thermal gradients, where the temperature increases with depth much faster than normal. During metamorphism, rocks behave plastically and are able to bend or stretch due to the high temperatures and pressures from the overlying strata. For this reason, metamorphic rocks make up the largest constituent of the Earth's crust (Table 5).

3

SEDIMENTARY STRATA

Most of the Earth's surface is covered by a thin veneer of sediment, and sedimentary rocks are encountered more frequently than any other type. They not only provide impressive scenery from ragged mountains to jagged canyons but also much of the wealth of the world, including valuable ores and petroleum. The sedimentary environment furnishes the conditions necessary for the formation of fossils, which provide important clues about the history of the Earth. The constant shifting of sediments on the Earth's surface and the accumulation of deposits on the ocean floor ensures that the face of the Earth will continue to change as time goes on.

SEDIMENTARY PROCESSES

Sediments are derived from weathering of the crust, and most sedimentary processes take place very slowly on the ocean bottom. Marine sediments consist of material washed off of the continents. Therefore, the majority of sedimentary rocks form along continental margins and in the basins of inland seas. Such a sea invaded the interior of North America during the Jurassic and Cretaceous periods (Figure 26). Areas with high sedimentation rates form deposits thousands of feet thick. Where these deposits are exposed on the surface, individual sedimentary beds can be traced for up to hundreds of miles.

The formation of sedimentary rock begins with erosion. The continents are the primary sites of erosion, whereas the oceans are primarily the sites of sedimentation. Rocks weather by several processes, including the action of wind, rain and ice. Loose sediment grains are carried downstream to the ocean. Rivers like the Amazon and the Mississippi transport enormous quantities of sediment derived from the interiors of the continents. The towering landform resulting from the collision of India and Asia is the greatest single source of sediment today. The sediment is hauled out by major rivers draining the region and emptied into the Bay of Bengal. This accounts for 40 percent of the total amount of sediment discharged into the ocean by all the rivers in the world.

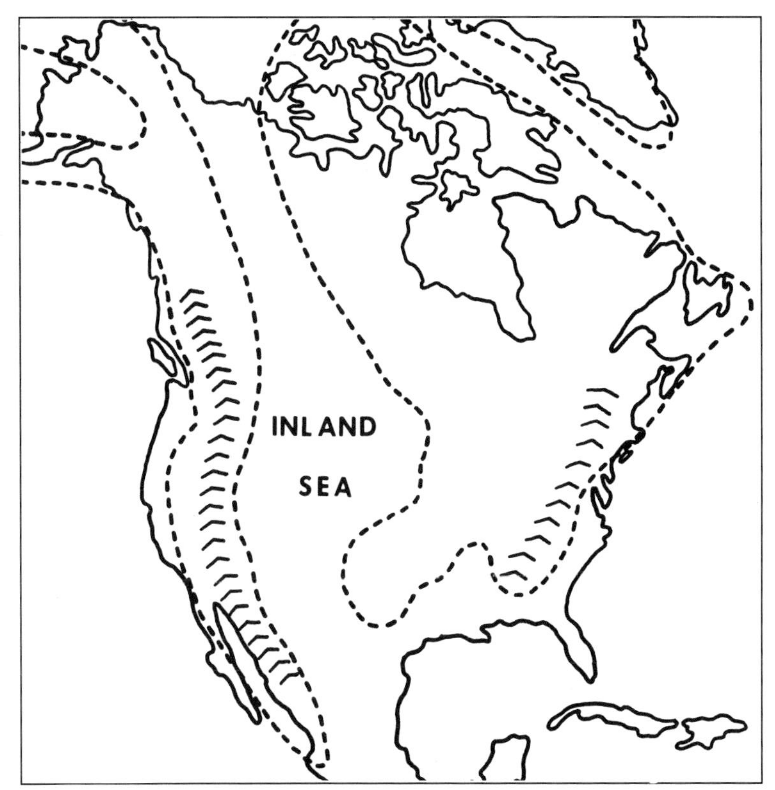

Figure 26 The Cretaceous interior sea of North America, where thick deposits of sediment were laid down.

Each year, an estimated 25 billion tons of sediment are carried by stream runoff into the ocean, where it settles onto the continental shelf. The continental shelf extends to 100 miles or more and is roughly 600 feet deep. In most places, the continental shelf is nearly flat with an average slope of only about 10 feet per mile.

Beyond the continental shelf lies the continental slope, which extends to an average depth of over 2 miles. It has a steep angle of several degrees, comparable to the slopes of many mountain ranges. Sediments reaching the edge of the continental shelf slide down the continental slope under the influence of gravity. Often, huge masses of sediment cascade down the continental slope by gravity slides, sometimes gouging out steep submarine canyons. A modern slide that broke submarine cables near Grand Banks, south of Newfoundland, moved downslope at about 50 miles per hour.

Figure 27 A massive dust storm in the western plains of the United States. Courtesy of USDA—Soil Conservation Service

Most minerals found in sedimentary rocks were precipitated directly from seawater. When land is eroded, some 3 billion tons of rock each year are dissolved in water and carried by streams to the sea. This is sufficient to lower the entire land surface of the Earth by as much as an inch in only 2,000 years. It is also one of the reasons why the ocean is so salty. Besides ordinary table salt, seawater contains large amounts of calcium carbonate, calcium sulfate and silica, which precipitate from seawater by chemical and biological processes.

In dry regions where dust storms are prevalent, loose sediment is carried by the wind (Figure 27). Windblown sediments landing in the ocean slowly build deposits of abyssal red clay, whose color signifies its terrestrial origin. However, most windblown sediments remain on the continents, where they accumulate into deposits of loess, which is distinguishable by its thin, uniform bedding. Most loess deposits in the central portion of the United States (Figure 28) were laid during the Pleistocene ice ages. During an ice age, regions not covered by glaciers dried out, causing widespread desertification.

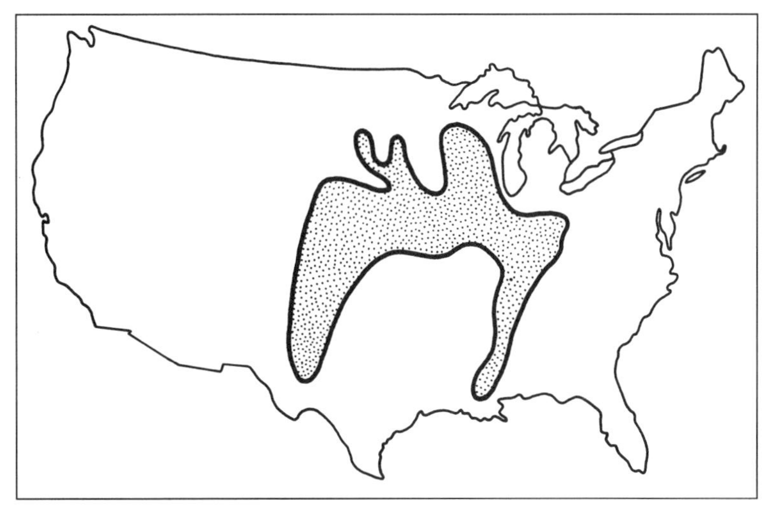

Figure 28 Location of windblown soil deposits in the United States.

Fluvial deposits are riverborne sediments that remain on the continents after erosion. When a river becomes clogged with sediments and fills its channel, it spills over onto the adjacent floodplain and carves out a new river course. This forces the river to meander along, forming thick sediment deposits in broad floodplains that can fill an entire valley. Fluvial deposits are recognized on outcrop by their coarse sediment grains and cross-bedding features (Figure 29), created when the stream meandered back and forth over old river channels.

Figure 29 Cross-bedding results when a stream meanders over old river channels.

Floodwaters rapidly flowing out of dry mountain regions carry a heavy sediment load, sometimes including blocks the size of automobiles. When the stream reaches the desert, the water rapidly percolates into the desert floor, causing the flood to suddenly halt. After erosion has peeled away the outer layers of mud, huge monoliths are left standing in the middle of nowhere as a testament to the power of moving water.

CLASTIC SEDIMENTS

Clastic sedimentary rocks are composed mainly of fragments broken loose from a parent material, deposited by mechanical transport and cemented into hard rock. The sediments are usually derived from the weathering or decomposition of igneous, metamorphic and other sedimentary rocks. The rocks are weathered, or broken down, into sediment grains by the action of wind, water, cycles of heating and freezing, and plant and animal activity. Weathering breaks rocks apart or causes their outer layers to peel or spall off in a process known as exfoliation.

The products of weathering include a wide range of materials from very fine grained sediments to large boulders. Erosion, either by wind, rain or glacial ice, eventually brings the sediments to streams and rivers, which in turn empty into the ocean. Angular sediment grains indicate a short time spent in transit, whereas rounded sediment grains indicate severe abrasion from long-distance travel or from reworkiing by fast-flowing streams or pounding ocean waves.

Solid rock exposed on the surface is chemically broken down into clays and carbonates and mechanically broken down into silts, sands and gravels. Streams heavily laden with sediments overflow their beds, forcing them to take several detours as they meander toward the sea. When the streams reach the ocean, their velocity falls sharply, and their sediment load drops out of suspension. Meanwhile, chemical solutions carried by the rivers are thoroughly mixed with seawater by ocean waves and currents.

When riverborne sediments reach the ocean, they settle out of suspension according to their grain size. The coarse-grained sediments settle out near the turbulent shore, and the fine-grained sediments settle out in calmer waters farther out to sea. As the shoreline advances toward the sea due to the buildup of coastal sediments or a diminishing sea level, finer sediments are progressively covered over by coarser ones. As the shoreline recedes due to the lowering of the land surface or a rising sea level, coarser sediments are covered by progressively finer ones. This produces a recurring sequence of sandstones, siltstones and shales (Figure 30). Gravels, however, are rare in the ocean and are mainly transported from the coast to the deep abyssal plains by submarine landslides called turbidity currents.

As the weight of the overlying sedimentary layers presses downward on the lower strata, the sediments are lithified into solid rock, providing a geologic column of alternating beds of limestone, shales, siltstones and sandstones. If these rocks are subjected to the heat and pressure of the Earth's interior, they are metamorphosed consecutively into marble, slate, quartzite and schist.

Shales and mudstones are the most abundant sedimentary rocks because they are the main weathering products of feldspars, which are the most abundant minerals. Furthermore, all rocks are eventually ground to clay-

Figure 30 Interbedded mudstone and sandstone formation, Capitol Reef National Monument, Utah.
Photo by J. R. Stacy, courtesy of USGS

size particles by abrasion. Because clay particles are so small and sink so slowly, they normally settle out in calm, deep waters far from shore. Compaction, resulting from the weight of the overlying sediments, squeezes out water between sediment grains, and the clay is lithified into mudstone if massive or shale if fissile, or thinly bedded.

Clastic sedimentary rocks are classified according to grain size. Gravel-size sediments are called conglomerates if rounded and breccias if angular. Conglomerates are composed of abundant quartz and chalcedony such as flint or chert. Breccias are relatively rare and are indicative of terrestrial mudflows or submarine landslides. Debris flows piled up on the continental slope can produce a coarse carbonate rubble known as brecciola. Volcanic breccias, also known as agglomerates, are consolidated pyroclastic fragments. Moraines and tillites are glacial deposits composed of a mixture of boulders and gravel-size sediments.

Sandstones are composed mostly of quartz grains roughly the size of beach sands. If a sandstone contains abundant feldspar, it is called an arkose. Graywacke, sometimes called dirty sandstone, is a dark, coarse-grained sandstone with a clay matrix and is believed to be deposited by submarine turbidity currents. Siltstones are composed of fine quartz grains that are just visible to the naked eye. Shales and mudstones are composed of the finest sedimentary particles, whose grains are not visible to the naked eye.

Clastic sediments are lithified into solid rock mainly by compaction with fine-grained sediments and cementation with coarse-grained sediments. With increasing weight from the overlying sedimentary layers, individual grains are compacted and solidified. Minerals such as calcium carbonate, silica or iron oxide are deposited between coarse sediment grains and are used as cementing agents.

Figure 31 The Big Elk coal bed in King County, Washington. Photo by J. D. Vine, courtesy of USGS

Coal beds (Figure 31) are considered sedimentary rock, even though they are not derived from clastic sediments or chemical precipitation. Coal originated from compacted plant material that once grew in lush swamps. Often between easily separated layers of coal or associated fine-grained sedimentary beds are carbonized remains of ancient plant stems and leaves. Black or carbonized shales also originated in the ancient coal swamps, and traces of plant life can be found between easily split shale layers.

CARBONATE ROCKS

Nonclastic, or precipitate, rocks are formed by biological or chemical precipitation of minerals dissolved in water. The term *precipitate* is actually a historical misnomer carried over from the days when these rocks were thought to form in ways similar to the precipitation of ice crystals. Rainwater contains a small amount of carbonic acid from the chemical reaction of atmospheric water vapor and carbon dioxide. The carbonic acid dissolves calcium and silica minerals in surface rocks to form bicarbonates. The bicarbonates are transported to the ocean by rivers and are thoroughly mixed with seawater by waves and currents.

The bicarbonates precipitate out of seawater by direct chemical processes or by biological activity, the two most common means. Organisms use calcium bicarbonate to build supporting structures such as shells composed of calcium carbonate. When the organisms die, their shells fall to the bottom of the ocean, where thick deposits of calcium carbonate slowly build up to form limestone.

Limestone is the most common precipitate rock and is mostly produced by biological activity. This is evidenced by abundant fossilized marine life in limestone beds. Some limestones are chemically precipitated in evaporite deposits. Dolomite, which resembles limestone, is created when the calcium in limestone is replaced by magnesium. It is more resistant to erosion from acid rain, which explains why the Dolomite Alps of Europe remain among the most impressive mountain ranges in the world.

Chalk is a soft, porous calcium carbonate rock and should not be confused with the chalk used on classroom blackboards, which is actually composed of calcium sulfate. Thick beds of chalk were deposited during the Cretaceous, which is how the period got its name from the Latin *creta* meaning "chalk." The soft chalk cliffs on the Suffolk Coast of England have been eroding away as much as 15 feet a year by the pounding waves of the North Sea. Sometimes a large storm erodes the tall cliffs several tens of feet.

Most limestones originated in the ocean, and some thin limestone beds were deposited in lakes and swamps. Limestones constitute approximately 10 percent of all exposed sedimentary rocks. Many limestones form mas-

Figure 32 The eastern Sawtooth Range in Lewis and Clark County, Montana is composed of Mississippian-age carbonates. Photo by M. R. Mudge, courtesy of USGS

sive formations (Figure 32). They are recognized by their typically light gray or light brown color. Whole or partial fossils constitute many limestones, depending on whether they were deposited in quiet or agitated waters.

The majority of carbonate sediments were deposited in fairly shallow waters, probably less than 50 feet deep, and mainly in intertidal zones, where marine organisms were plentiful. Coral reefs, which form in shallow water where sunlight can easily penetrate for photosynthesis, contain abundant organic remains. Many ancient carbonate reefs are composed largely of carbonate mud that contains large skeletal remains.

Most carbonate rocks began as sandy or muddy calcium carbonate material. The sand-sized particles are composed of broken-up skeletal remains of invertebrates and shells of calcareous algae that rain from above. The skeletal remains might have been broken up by mechanical means, such as the pounding of the surf, or by the activity of living organisms. Further breakdown into dust-size particles produces a carbonate mud, the most common constituent of carbonate rocks, and forms a matrix known as micrite. Under certain conditions, the carbonate mud dissolves in seawater and is redeposited elsewhere on the ocean floor, forming a calcite ooze that later lithifies into limestone.

As calcareous sediments accumulate into thick deposits on the ocean floor, deep burial of the lower strata produces high pressures, which lithifies the beds into carbonate rock, consisting mostly of limestone or dolostone. If fine-grained calcareous sediments are not strongly lithified, they form deposits of soft, porous chalk. Limestones typically develop a secondary crystalline texture, resulting from the growth of calcite crystals by solution and recrystallization following the formation of the original rock.

Some carbonate rocks were deposited in deep seas. The maximum depth carbonate rocks can form is determined by the calcium carbonate compensation zone, which generally begins at a depth of about 2 miles. Below this zone, the cold, high-pressure waters of the abyssal, which contain the vast majority of free carbon dioxide, dissolve calcium carbonate sinking to this level. The upwelling of deep-ocean water, mainly in the tropics, returns to the atmosphere carbon dioxide lost by the carbon cycle, which is the circulation of carbon dioxide by geochemical processes (Figure 33).

Silica readily dissolves in seawater in volcanically active areas on the seafloor, from volcanic eruptions into the sea, and from weathering of siliceous rocks on the continents. Some organisms like sponges and diatoms extract the dissolved silica directly from seawater to build their shells and skeletons. Accumulations of siliceous sediment on the ocean floor

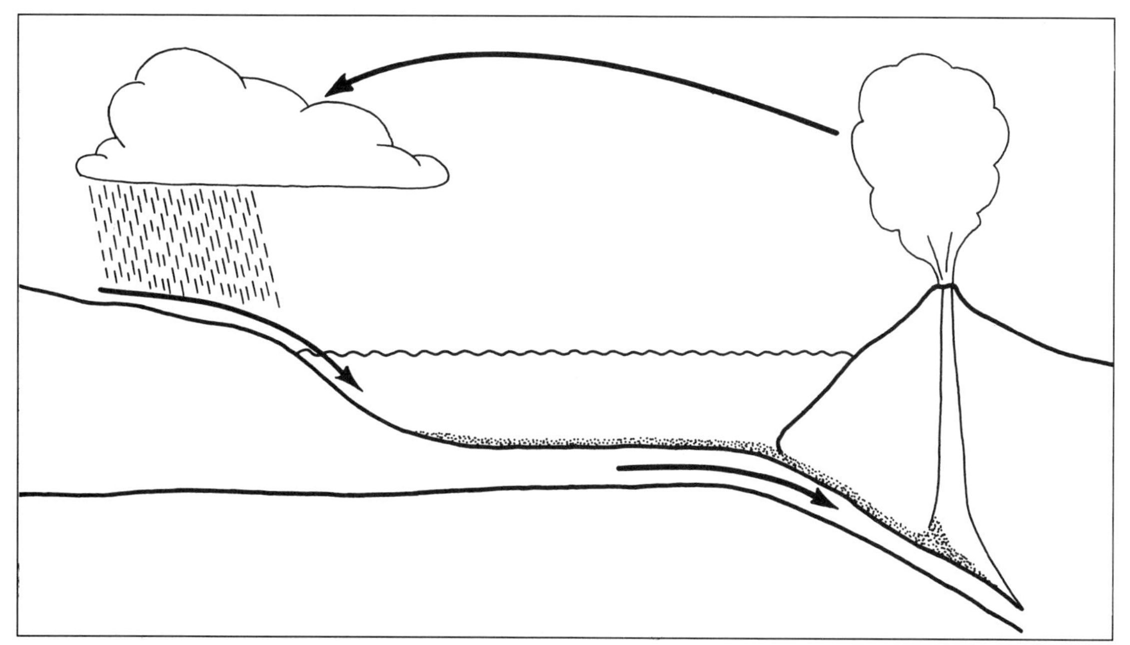

Figure 33 The carbon cycle. Carbon dioxide in the form of bicarbonate is washed off the land and enters the ocean where organisms convert it into carbonate sediments, which are thrust into the mantle, become part of the magma and escape into the atmosphere from volcanoes.

from dead organisms form diatomaceous earth also called diatomite. Thick deposits throughout the world are a tribute to the prodigious growth of these organisms during the last 600 million years.

EVAPORITE DEPOSITS

Seawater contains about 3.5 percent dissolved minerals, mostly sodium chloride, or common salt. Salt has been mined extensively throughout the world since ancient times. It is extracted from formerly shallow, stagnant pools of seawater called brines that have evaporated, which is why they are called evaporite deposits. The evaporation occurs in shallow, slowly sinking basins that are partially dammed by sandbars. During storms, seawater flows over the sandbars and replenishes the basins. Salt also accumulates in thick beds in deep basins that were cut off from the general circulation of the ocean.

As seawater evaporates, the concentration of salt increases to the saturation point. The salt then precipitates out of solution and accumulates on the seafloor. Several thousand feet of seawater must evaporate in order to produce 100 feet of salt. But many salt deposits are much thicker than this, possibly signifying many cycles of evaporation.

Several other minerals are also deposited, including gypsum, which is used in wallboard; phosphates and nitrates, used in fertilizers and explosives; potassium, used in a wide variety of products such as fertilizers; and important halogens such as bromine, chlorine and iodine. Evaporite deposits in the interiors of continents, such as the potassium deposits at Carlsbad, New Mexico, indicate that these areas were once inundated by the sea. Many oil fields are located near ancient salt domes, which provide the structures needed for trapping oil and gas.

Evaporite deposits generally form under arid conditions between 30 degrees north and 30 degrees south of the equator. However, extensive salt deposits are not being formed at the present time, which suggests a cooler global climate. That ancient evaporite deposits exist as far north as the arctic regions indicates that either these areas were at one time closer to the equator or the global climate was considerably warmer in the geologic past. Evaporite accumulation peaked about 230 million years ago, when the supercontinent Pangaea was beginning to rift apart. Few evaporite deposits date beyond 800 million years ago, however, probably because most of the salt formed before then has been recycled.

Presently, the Mediterranean Sea is practically an enclosed basin (Figures 34A and 34B). The evaporation rate is very high, and nearly 5 feet of the water's surface evaporate every year. This generates water with a high salt content, which makes it heavier than normal seawater. The highly saline water sinks to the bottom and will eventually fill the entire basin.

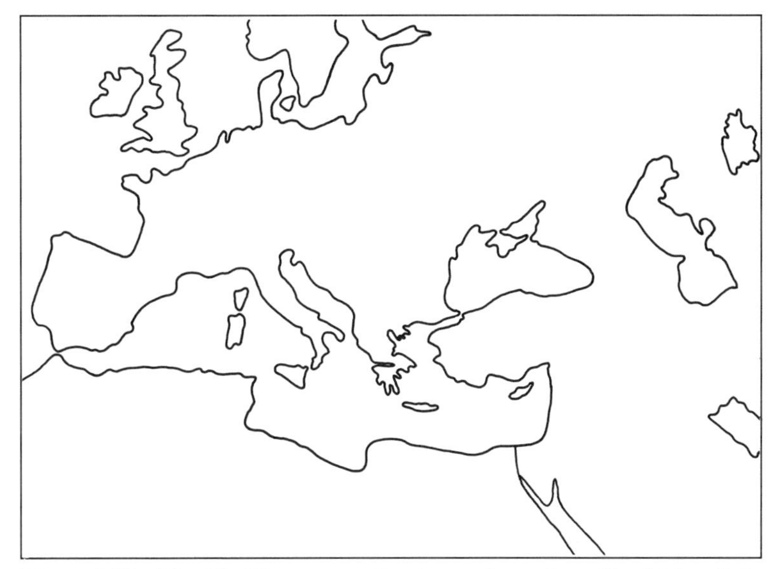

Figure 34A The Mediterranean Sea is nearly enclosed by Africa, Asia and Europe.

Futhermore, the inflow from rivers into the Mediterranean cannot compensate for the evaporation and the outflow of water at the Gibraltar shelf.

Six million years ago, the Mediterranean Basin was completely cut off from the Atlantic Ocean when Gibraltar was uplifted, forming a dam across the strait. Over about 1,000 years the entire sea, amounting to nearly 1 million cubic miles of water, evaporated, leaving behind a dry basin over 1 mile deep. On the bottom, salt deposits formed as the salt content of the water column precipitated and was deposited on the basin floor. Rivers draining into the desiccated basin gouged out deep canyons. Burried under the sediments of the Nile Delta is a 1 mile-deep canyon comparable in size to the Grand Canyon.

After about 1 million years, Gibraltar subsided and the dam was broken. This created a spectacular waterfall, disgoring water at a rate of 10,000 cubic miles a year, 1,000 times greater than Niagara Falls. Subsequent sedimentation buried trillions upon trillions of tons of salt beneath younger sediments. The Gulf of Mexico also completely dried out 140 million years ago. The North Sea and the Red Sea might also have experienced a similar fate; they show signs of evaporation by being floored with thick layers of salt.

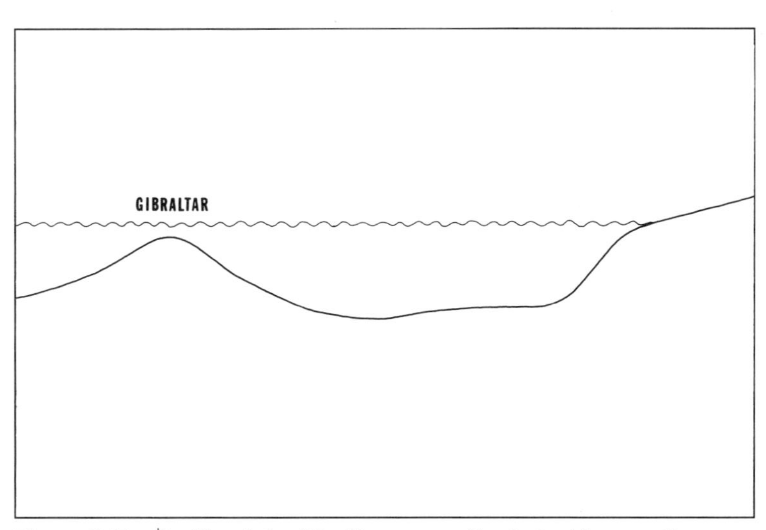

Figure 34B Profile of the Mediterranean Basin looking north.

The salts precipitate out of solution in stages. The first mineral to precipitate is calcite, closely followed by dolomite, although only minor amounts of limestone and dolostone are deposited in this manner. After about two thirds of the water has evaporated, gypsum precipitates. When nine tenths of the water is removed, halite, or common salt, forms. Thick deposits of halite are also produced by the direct precipitation of seawater in deep basins that have been cut off from the ocean.

Thick beds of gypsum, composed of hydrous calcium sulfate, constitute one of the most common sedimentary rocks. They are produced in evaporite deposits that formed when a pinched-off portion of the ocean or an inland sea evaporated. Oklahoma, like many parts of the interior of North America that were invaded by a Mesozoic sea, is well known for its gypsum beds. The mineral is mined extensively for the manufacture of plaster, which was first used by the ancients in Asia for flooring material, containers, sculptures and ornamental beads as early as 12,000 B.C., long before the invention of pottery.

SEDIMENTARY STRUCTURES

Layers of sedimentary rock are separated by bedding planes, which are areas of weakness where the rocks tend to separate or break apart. The varying thicknesses of the layers reflects different depositional environments at the time the sediments were laid. Each bedding plane generally marks where one type of deposit ends and another begins. Thus, thick sandstone beds might be interspersed with thin beds of shale, indicating that periods of coarse sediment deposition were punctuated by periods of fine sediment deposition brought on possibly by changing climate conditions.

Graded bedding occurs when the particles in a sedimentary bed vary from coarse at the bottom to fine

Figure 35 Ripple marks on Dakota Sandstone, Jefferson County, Colorado. Photo by J. R. Stacy, courtesy of National Park Service

at the top. This type of bedding indicates rapid deposition of sediments of differing sizes by a fast-flowing stream emptying into the sea. The largest particles settle out first and are covered with progressively finer material due to the difference in settling rates. Beds can also grade laterally, producing a gradation of sediments called a facies change, from the Latin word for "form."

The color of sedimentary beds helps to identify the type of depositional environment. Generally, sediments tinted various shades of red and brown indicate a terrestrial source, whereas green and gray sediments suggest a marine environment. The sizes of individual particles also influence the color intensity, with darker-colored sediments generally indicating finer grains.

Fluvial, or river, deposits are recognized in outcrops by their course sediment grains and cross-bedding features, generated when a stream meanders back and forth over old river channels. River currents can also align mineral grains and fossils, giving rocks a linear structure that can be used to determine the direction of current flow. Ripple marks on exposed surfaces (Figure 35) can also be used to determine direction of current flow or the wind direction if the surfaces are composed of desert sand.

Most windblown sediments form thick layers of loess, a fine-grained, sheetlike deposit, which on outcrop might show beds of uniform thickness.

TABLE 6 MAJOR DESERTS OF THE WORLD

Desert	Location	Type	Area (square miles × 1,000)
Sahara	North Africa	Tropical	3,500
Australian	Western/interior	Tropical	1,300
Arabian	Arabian Peninsula	Tropical	1,000
Turkestan	S. Central U.S.S.R.	Continental	750
North American	S.W. U.S./N. Mexico	Continental	500
Patagonian	Argentina	Continental	260
Thar	India/Pakistan	Tropical	230
Kalahari	S.W. Africa	Littoral	220
Gobi	Mongolia/China	Continental	200
Takla Makan	Sinkiang, China	Continental	200
Iranian	Iran/Afghanistan	Tropical	150
Atacama	Peru/Chile	Littoral	140

Figure 36 Large dunes in Death Valley, California. Courtesy of National Park Service

Along with these are dune deposits, composed of desert sand, which when lithified show a distinct dune structure on outcrops. The dunes move across the desert floor in response to the wind by a process known as saltation, whereby sand grains in motion dislodge each other and become airborne for a moment. In this manner, sand dunes engulf everything in their paths, including some man-made structures as they march across the desert floor (Figure 36). The sediment grains of desert deposits are often frosted due to the constant motion of the sand, which causes abrasion. Rocks are often polished by wind abrasion similar to sandblasting or coated with a desert varnish from mineral solutions exuded from within.

4

EROSIONAL PROCESSES

Erosion is a natural geologic process that cuts down tall mountains and carves deep canyons and has been doing so since the very beginning of time. No matter how pervasive the formation of mountain ranges by the forces of uplift is, the mountains eventually lose the battle with erosion and are worn to the level of the prevailing plain. Erosion also gouges deep ravines in the hardest rock and has obliterated most geologic structures, including almost all large meteorite craters. Over a few thousand years, erosion has eradicated almost all man-made structures from ancient civilizations.

THE HYDROLOGIC CYCLE

The movement of water on the planet, known as the hydrologic cycle (Figure 37), is one of nature's most important cycles. The oceans cover about 70 percent of the Earth's surface with an average depth of over 2 miles, amounting to nearly 250 million cubic miles of water. Three percent of the Earth's water is fresh, enough to fill the Mediterranean Sea 10 times over. About three quarters of all freshwater is locked in glacial ice. The remainder is atmospheric water vapor, running water in rivers, stand-

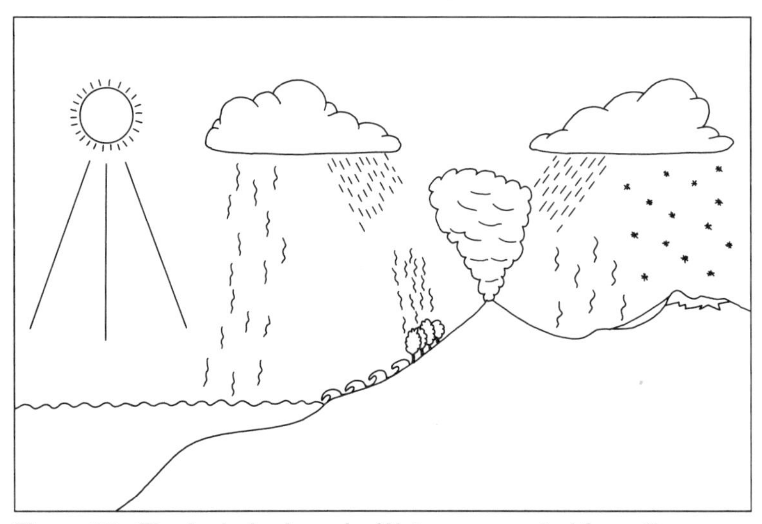

Figure 37 The hydrologic cycle. Water evaporated from the ocean precipitates on the land, erodes the landscape and returns to the ocean.

ing water in lakes, groundwater, soil moisture and water in plant and animal tissues.

Each year, the sun evaporates 100,000 cubic miles of seawater, which returns to Earth in the form of precipitation, mostly directly back into the ocean. About 10 percent, or some 10,000 cubic miles of water, falls on the continents and eventually returns to the sea. Roughly two thirds of the rain that falls evaporates either directly or by the transpiration of plants. Uneven distribution of rainfall results in dry deserts and wet tropical rain forests. Insufficient rainfall causes droughts, and excessive rainfall causes floods.

The average journey water takes from the ocean to the atmosphere, over the land, and back to the sea is about 10 days. The journey is only a few hours long in the tropical coastal regions and as much as 10,000 years in the polar regions. The quickest route water can take back to the ocean is by runoff in rivers. This is perhaps the most important part of the hydrologic cycle. As the surface water makes its way to the ocean, it transports sediments washed off the continents, ultimately eroding them to the level of the sea.

EFFECTS OF EROSION

Erosion rates vary depending on the amount of precipitation, the topography of the land, the type of rock and soil material, and the amount of vegetative cover. Each year the Mississippi River dumps more than 250 billion tons of sediment into the Gulf of Mexico, widening the Mississippi Delta and slowly building up Louisiana and nearby states. The Gulf Coast states from East Texas to the Florida panhandle were built up by sediments eroded from the interior of the continent and hauled along by the Mississippi and other rivers. The Imperial Valley of southern California owes its rich soil to the Colorado River, which carved out the mile-deep Grand Canyon and deposited its sediments 3 miles thick in the region.

The process of erosion is delicately balanced by buoyancy, which keeps the continents afloat. Erosion can only shave off the top portion of the continental crust before the mean height of the crust falls below sea level, at which point erosion ceases and sedimentation commences. In the past, erosion rates were probably higher than they are today. In the Earth's early stages, the relief of the land was not nearly as great as it is now. It took eons of mountain building and erosion to give us our present landscape of tall mountains and deep canyons.

The rise of active mountain chains such as the Himalayas (Figure 38) is matched by erosion so that their net growth is almost zero. The cores of the world's mountain ranges contain some of the oldest rocks. What was once buried deep in the bowels of the Earth is now thrust high above. Huge blocks

Figure 38 The Himalaya Mountains of India and China viewed from the space shuttle. Courtesy of NASA

Figure 39 Severe erosion on farmland near Viola, Idaho, showing gully and rills. Photo by Carrol Tyler, courtesy of USDA—Soil Conservation Service

of granite that formed the interiors of the continents were pushed up by tectonic forces operating deep within the Earth and exposed by erosion.

Soil erosion causes the most widespread degradation of the land surface. Falling rain erodes surface material by impact and runoff. The impact of raindrops striking the ground with a high velocity loosens material and splashes it up into the air. On hillsides, some of this material falls back at a point lower down the slope. About 90 percent of the energy is dissipated by the impact. Most of the impact splashes are up to 1 foot high, and the lateral splash movement is about four times the height.

Impact erosion is most effective in regions with little or no vegetative cover and that are subjected to sudden downpours, such as desert areas. Splash erosion accounts for the puzzling removal of soil from hilltops where there is little runoff. It could also ruin soil by splashing up the light clay particles, which are carried away by runoff, leaving infertile silt and sand behind. Rainwater not infiltrating into the ground runs down the hillside and erodes the soil, cutting deep gullies into the terrain (Figure 39). The degree of erosion depends on the steepness of the slope and the type and amount of vegetative cover.

Figure 40 This sea cliff at Moss Beach, San Mateo County, California has receded 165 feet in 105 years since 1866. Courtesy of USGS

Breaking waves dissipate energy along the coast and are responsible for generating along-shore currents, which transport sand along the beach. They also cause coastal erosion, a serious problem in many areas where the shoreline is steadily receding (Figure 40). Most of the high waves and beach erosion occurs during coastal storms. Hurricanes with winds of 100 miles per hour and more produce the most dramatic storm surges, which are responsible for destroying entire beaches. The process of beach erosion is mainly influenced by the strength of beach dunes or sea cliffs, the intensity and frequency of coastal storms and the exposure of the coast.

As beaches along the East Coast erode, natural processes will not replenish the disappearing sand until the next ice age. Most of the sand along the coast and continental shelf originates in the north from sources such as the Hudson River. For the sand to move as far south as the Carolina coast, it had to progress in steps that possibly took millions of years. As the sand moves along a coast, ocean currents push it into large bays or estuaries. The embayment will continue to fill with sand until sea levels drop and the accumulated sediment is flushed down onto the continental shelf. The sand can travel only as far as the next bay in a single glacial cycle, however. Therefore, most beaches will not receive a major restocking of sand until the next ice age.

LANDSLIDES

Landslides are violent shifts of earth materials resulting mainly from earthquakes and weather systems and are among the most powerful erosional agents both on land and under the sea. They are the mass movement of soil and rock material downslope under the influence of gravity. Cali-

Figure 41 Sliding of unstable earth materials undermines the foundations of homes in the Pacific Palisades area of southern California. Photo by J. T. McGill, courtesy of USGS

fornians are well aware of frequent landslides in their state (Figure 41). During the last decade, there were thousands of landslides in the Los Angeles Basin alone.

All slides result from the failure of earth materials under stress. They are initiated by an increase in stress and a reduction of strength. The addition of water to a slope can contribute to an increase in stress and a decrease in strength. Slides are also caused by the removal of lateral support by stream, glacier or wave erosion as well as longshore or tidal currents. The loss of shear strength depends on the composition, texture and structure of the soil and the slope geometry. Changing water content and pore pressure also act like lubricants between rock layers.

Landslides are rapid movements of overburden with or without the underlying bedrock. Slides consisting of overburden alone are called debris slides. Slides involving bedrock are called rockslides and slump. Rockslides develop when a mass of bedrock is broken into many fragments during the fall, and the material behaves like a fluid, spreading out on the valley below (Figure 42). With enough velocity, it might even flow some distance uphill on the other side of the valley. Such landslides are called avalanches, although this term is generally used to describe snowslides (Figure 43). Rockslides are usually large and destructive, often involving millions of tons of rock. They develop when planes of weakness, such as bedding planes or jointing, are parallel to a slope, especially if the slope has been undercut by erosion.

Slumps develop where a strong, resistant rock overlies weak rocks. The material slides in a curved plane, tilting up the resistant unit, while the weaker rock flows out to form a heap. Unlike rockslides, slumps develop new cliffs nearly as high as those previous to the slumping, which sets the stage for a new slump. Slumping is therefore a continuous process, and generally many previous generations of slumps can be seen far in front of the present cliffs.

Material that drops at nearly the velocity of free-fall from a near vertical mountain face is called a rockfall or soilfall, depending on its composition. Rockfalls can be individual blocks dropping down a mountain slope or masses weighing hundreds of thousands of tons falling nearly straight down the mountain face. Individual blocks commonly come to rest in a loose pile of angular blocks, called talus, at the base of a cliff (Figure 44). If large blocks of rock drop into a body of water, immensely destructive waves called tsunamis are set in motion. Coastal and submarine slides due to earthquakes can also cause destructive tsunamis.

Coastal landslides are produced when a sea cliff is undercut by wave erosion and falls into the ocean. Excessive rainfall along the coast can also lubricate sediments, allowing large blocks to slide into the sea. Submarine slides move down steep continental slopes and have been known to bury transcontinental telephone cables under thick layers of sediment. Submarine slides

Figure 42 A 700-ton boulder transported by a landslide from the May 31, 1970, Peruvian earthquake. Courtesy of USGS

might also be responsible for carving deep canyons in the ocean floor. The slides consist of sediment-laden water, and because they are more dense than seawater they move rapidly along the ocean floor and can severely erode the soft bottom material.

MASS WASTING

Other types of earth movements go under the general heading of mass wasting. This is the mass transfer of earth material, resulting in slipping, sliding and creeping down even the gentlest of slopes, and it is directly influenced by the amount of soil water. Creep is a slow downslope movement of soil (Figure 45) and is recognized by downhill-pointing poles, fence posts and trees. Creep might be more rapid in areas where freeze-thaw cycles cause material to move downslope due to the expansion and contraction of the soil.

Figure 43 A slab avalanche triggered by a skier. Courtesy of USGS

A rise in the soil water content increases weight and reduces stability by lowering the resistance to shear, resulting in an earthflow. This is a transition between the slow and rapid varieties of mass wasting and is a more visible form of movement. Earthflows usually have a spoon-shaped sliding surface, whereupon a tongue of soil breaks away and flows a short distance, leaving a large curved scarp at the breakaway point.

With a further increase in water content, an earthflow grades into a mudflow. This is a highly viscous fluid, which often carries a tumbling mass of rocks and boulders, some the size of automobiles.

Figure 44 Talus cones in Stinking Water Canyon, Park County, Wyoming. Courtesy of USGS

Mudflows are the most impressive feature of many of the world's deserts. Heavy rainfall in bordering mountain regions causes sheets of water to run rapidly down steep mountain slopes, picking up large amounts of material on their way. This produces a swift flood of muddy material often with a steep wall-like front. Such a mudflow can cause serious damage as it flows out of the mountains.

Mudflows arising from volcanic eruptions are called lahars. Lahars are masses of water-saturated rock debris that move down the slopes of a volcano in a manner resembling the flowage of wet concrete. A lahar from the 1985 eruption of Nevado del Ruiz killed 25,000 people in Armero, Columbia. The debris is commonly derived from masses of loose unstable rock deposited on the flanks of a volcano by explosive eruptions. The water is provided by rain, melting snow, a crater lake, or a lake or reservoir adjacent to the volcano. Lahars can also be induced by a pyroclastic or lava flow moving across a snowfield, causing it to rapidly melt.

Figure 45 Railroad tracks damaged by soil creep near Coal Creek, Canada. Photo by W. W. Atwood, courtesy of USGS

Figure 46 Sand boil on Ten Mile Hill from the August 31, 1886, Charleston, South Carolina, earthquake. Photo by J. K. Hillers, courtesy of USGS

Liquefaction is responsible for producing ground failure during earthquakes and violent volcanic eruptions. It is restricted to certain geologic and hydrologic environments, mainly areas where sands and silts were deposited possibly within the last 10,000 years and where the groundwater table is close to the surface. Generally, the younger and looser the sediments and the higher the water table the more susceptible the soil is to liquefaction.

Ground failures associated with liquefaction are lateral spreads, flow failures and loss of bearing strength. In addition, liquefaction enhances ground settlement or subsidence. Sand boils (Figure 46), which are fountains of water and sediment that spout from the pressurized liquefied zone, can reach 100 feet or more high. Sand boils are produced when water laden with sediment is vented to the surface by artesianlike water pressures developed during the liquefaction process. Sand boils can also cause local flooding and accumulations of large amounts of silt and sand.

Lateral spreads are the lateral movement of large

blocks of soil resulting from liquefaction in a subsurface layer caused by earthquakes. Lateral spreads generally develop on gentle slopes, most commonly less than 6 percent. Horizontal movements on lateral spreads are as much as 10 to 15 feet. But where slopes are particularly favorable and the duration of the earthquake is long, lateral movement might be up to 10 times greater. Lateral spreads usually break up internally, forming numerous fissures and scarps, but seldom is the damage catastrophic.

Figure 47 Landslide in Anchorage, Alaska, following the March 27, 1964, earthquake. Photo by W. R. Hansen, courtesy of USGS

The most catastrophic type of ground failure due to liquefaction is flow failure (Figure 47), consisting of liquefied soil or blocks of intact material riding on a layer of liquefied soil. These failures usually move up to 100 feet or more, but under certain geographical conditions they can travel several miles at speeds of up to 30 or more miles per hour. Flow failures usually form in loose, saturated sands or silts on slopes greater than 6 percent and originate on land and on the seafloor.

FLOODS

Flash floods are the most intense form of flooding. They are local floods of great volume and short duration and are generally caused by torrential rains or cloudbursts associated with severe thunderstorms on a relatively small drainage area. Flash floods also result from a dam break or from a sudden breakup of an ice jam, causing the release of a large volume of flow in a short time. A special type of flash flood occurred during the 1980 eruption of Mount St. Helens, which created major mudflows and flooding from melted glaciers and snow on the volcano's flanks. Flood-hazard zones extend considerable distances down some valleys. Volcanoes in the western Cascade Range, for example, have flood zones that can reach as far as the Pacific Ocean.

Violent thunderstorms can produce flash floods on widely dispersed streams, resulting in high flood waves. The discharges quickly reach a maximum and diminish almost as rapidly. Floodwaters frequently contain large quantities of sediment and debris collected as they sweep clean the stream channel. Flash floods can take place in almost any part of the country, but they are especially common in the mountainous areas and desert regions of the West. They are particularly dangerous in areas where the terrain is steep, surface runoff rates are high, streams flow in narrow canyons and severe thunderstorms are commonplace (Figure 48).

Riverine floods are produced by precipitation over large areas, by the melting of the winter's accumulation of snow, or both. They differ from flash floods in both extent and duration and take place in river systems whose tributaries drain large geographic areas and encompass many independent river basins. Floods on large river systems might last from a few hours to many days. The flooding is influenced primarily by variations in the intensity, amount and distribution of precipitation. Other factors directly affecting flood runoff are the condition of the ground, the amount of soil moisture and the vegetative cover.

River channel storage, changing channel capacity, which depend on the river size, and timing of flood waves control the movement of floodwaters.

Figure 48 Destruction of Drake, Colorado, from the July 31, 1976, Big Tompson River flood. Courtesy of USGS

As the flood moves down the river system, temporary storage in the channel reduces the flood peak. As tributaries enter the main stream, the river increases in size farther downstream. Since tributaries are not the same size or spaced uniformly, their flood peaks reach the main stream at different times, thereby smoothing out the peaks as the flood wave moves downstream.

Figure 49 Trellis and dendritic drainage patterns in the Utukok-Corwin region, northern Alaska. Photo by R. M. Chapman, courtesy of U.S. Air Force and USGS

DRAINAGE PATTERNS

About 3.5 million miles of rivers and streams cross the United States. Most of the water received on the continent is lost through floods or is held in lakes, swamps and soil. About a third is base flow, which is the stable runoff of rivers and streams. Another third is subsurface flow, which discharges mostly through evaporation, and only about 1 percent reaches the ocean. Groundwater travels very slowly and can make continental-scale journeys that take up to millions of years.

A drainage basin is the entire area from which a stream and its tributaries receive their water. For example, the Mississippi River and its tributaries drain a tremendous section of the central United States, reaching from the Rockies to the Appalachian Mountains. Moreover, each of its tributaries has its own drainage area, which forms a part of the larger basin. Individual streams and their valleys are joined together into networks, which display various types of drainage patterns, depending on the terrain.

Drainage patterns might be dendritic, resembling the branches of a tree, if the terrain is of uniform composition, or trellis, which is rectangular due to differences in the bedrock's resistance to erosion (Figure 49). Rectangular drainage patterns also occur if the bedrock is crisscrossed by fractures, which form zones of weakness that are particularly susceptible to erosion. Streams that radiate in all directions from a topographic high, such as a volcano, form radial stream patterns.

Stream drainage patterns, which are influenced by topographic relief and rock type, provide important clues about the type of geologic structure in an area. In addition, structure color and texture carry information about the rock formations it comprises. Surface expressions such as domes, anticlines, synclines and folds bear clues about the subsurface structure. Various types of drainage patterns infer variations in the surface lithology. The drainage pattern density is another good indicator of the lithology. Variations in the drainage density are also associated with variation in the coarseness of the alluvium.

In areas of exposed bedrock, these patterns depend on the lithologic character of the underlying rocks, the attitude of these rock bodies and the arrangement and spacing of planes of weakness encountered by runoff. Any abrupt changes in the drainage patterns are particularly important because they signify the boundary between two rock types, which might be good locations to explore for minerals.

EROSIONAL FEATURES

Some of the most impressive geologic features the planet has to offer were carved out of the crust by erosional processes. Massive sandstone cliffs that dwell in the western United States (Figure 50) have been slowly eroding for millions of years. Perhaps nowhere else on Earth is this process more illuminating than in the Grand Canyon of northern Arizona, which was carved out by the roaring Colorado River, now only a trickle of its former self.

Farther north is Bryce Canyon National Park in southern Utah, where fantastic pillars were carved out of the colorful Oligocene Wasatch Formation. Similarly colored sediments are responsible for the Painted Desert of Arizona and the Badlands of South Dakota (Figure 51), where short, steep slopes were eroded by numerous small streams, forming a unique drainage network.

Erosional processes taken to extreme have created Monument Valley on the border between Utah and Arizona. Isolated, or groups of, monuments rise 1,000 feet or more off the desert floor. A resistant cap rock preserved the sediments below, and the rest of the landscape was eroded. A similar

Figure 50 Sandstone cliffs of late Cretaceous age in Mesa Verde National Park, Colorado. Courtesy of National Park Service

Figure 51 Rugged outcrops of the Oligocene Wasatch Formation, Badlands National Park, South Dakota. Photo by J. J. Palmer, courtesy of National Park Service

situation only on a broader scale exists in flattop mountains, where a remnant of the original peneplain, which literally means almost a plain, is protected by a more resistant layer of sandstone. Many mesas, such as the Grand Mesa in western Colorado, the largest in the world, owe their existence to an upper layer of resistant basalt.

Dikes, formed by tabular magma bodies occupying a crack or a fissure in the crust, are usually harder than the surrounding material, and form long ridges when exposed by erosion. One of the best examples of this feature is Shiprock, New Mexico, where large dikes radiate from a 1,400-foot volcanic neck, left standing when the overlying sediments were eroded. Another good example of this type of structure is Devil's Tower in Wyoming, which is composed of solidified magma that filled a volcanic pipe, and erosion has left the more resistant rock standing high above surrounding terrain.

5

TYPE SECTIONS

The Earth's numerous geologic layers, or stratigraphic units, have a complex system of classification. Stratigraphic units are classified into erathems, consisting of the rocks formed during an *era* of geologic time. Erathems are divided into systems, consisting of rocks formed during a *period* of geologic time. Systems are divided into groups, consisting of rocks of two or more formations that contain common features. Formations are classified by distinctive features in the rock and are given the name of the locality where they were originally described. Formations are subdivided into members, which might be further divided into individual beds such as sandstone, shale or limestone.

A type section is a sequence of strata that was originally described as constituting a stratigraphic unit and serves as a standard of comparison for identifying similar widely separated units. Preferably, a type section is selected in an area where both the top and bottom of the formation are exposed. Type sections are named for the area where they are best exposed. For example, the Jurassic Morrison Formation, which is well known for its dinosaur bones (Figure 52), is named for the town of Morrison near Denver, Colorado.

Type sections are also distinguished by their distinct fossil content, which is used to correlate stratigraphic units. These are placed in order by age into a geologic column and are used to establish a geologic time scale.

Figure 52 The Jurassic Morrison Formation, Uinta Mountains, Utah.

Dated material is used to place actual ages on units of geologic time. Finally, a geologic map presents in plan view the geologic history of an area where particular rock exposures are found.

GEOLOGIC TIME

The history of the Earth has been divided into units of geologic time according to the type and abundance of fossils in the strata. The major units were delineated during the nineteenth century by geologists in Great Britain and western Europe. The periods take their names from the localities that have the best exposures. For example, the Jurassic period was named for the Jura Mountains in Switzerland.

Since there was no means of determining the actual ages of rocks during this time, the entire geologic record was created by using relative dating techniques, which places geologic time units in their proper sequence without reference to their actual age. It was only during this century that absolute dates have been added to units of geologic time after the development of radiometric dating techniques, based on the decay of radioactive isotopes.

Geologic time is divided into intervals called eras. They include the Precambrian, known as the age of prelife, the Paleozoic, known as the age of ancient life, the Mesozoic, known as the age of middle life, and the Cenozoic, known as the age of new life. The largest era is the Precambrian, which is 4 billion years in duration, and mostly obscure owing to the scarcity of fossilized remains of ancient organisms. It was not until the beginning of the Paleozoic era, about 570 million years ago, that the fossil record vastly improved due to the proliferation of species with hard skeletons.

The eras following the Precambrian are divided into smaller units called periods. There are seven periods in the Paleozoic, three in the Mesozoic and two in the Cenozoic. Each period is characterized by changes in the organisms that are less profound compared to the eras, which mark the boundaries of mass extinctions, proliferations or rapid transformations of

Figure 53 Cambrian-age sandstone at Pictured Rocks, Michigan, along Lake Superior. Courtesy of National Park Service

Figure 54 A meteorite found in Antarctica is thought to be of possible Martian origin. Courtesy of NASA

species. The two periods of the Cenozoic have been further subdivided into seven epochs because of the greater detail provided by recent rocks. The epoch we live in is called the Holocene, which corresponds to the neolithic in archaeology and the beginning of civilization.

The lack of fossils in ancient rocks often puzzled early geologists. Then suddenly life appeared in rocks at the base of the Cambrian period in great abundance at the same horizon the world over (Figure 53). The Cambrian period was named for the Cambrian mountain range in central Wales, where sediments containing the earliest known fossils were found. The base of the Cambrian was thought to be the beginning of life, and all time before then was simply called Precambrian.

AGE OF THE EARTH

The age of the Earth can be compared with the length of a single day. Half an hour after midnight, the Earth emerged from a collection of primordial dust and gas. Life first appeared about 3:00 A.M. By 4:00 P.M. the first single-celled animals evolved. Multicellular animals called metazoans arrived around 8:00 P.M. The first vertebrates followed 1 hour later and conquered the land less than an hour after that. The dinosaurs turned up about 11:00 P.M. They were supplanted by the mammals about half an hour later. One minute to midnight man came along.

The sixth-century B.C. Greek philosopher Xenophanes was perhaps the first to speculate on the age of the Earth. He believed that rocks containing fossil seashells in the mountainsides were evidence that they must have originated in the sea below. He thought that the Earth had to be extremely old to allow enough time for the imperceptible growth of mountains. Otherwise, some spectacular catastrophe would have occurred that rapidly raised them high above the sea.

Early geologists tried a number of methods to determine the age of the Earth. One was to measure the thickness of sedimentary strata and compare it to known sedimentation rates. This method was notoriously unreliable and provided a variety of dates, depending on which locality was used because some deposits were much thicker than others. Another method was to compare the salinity of rivers to that of the ocean, which was thought to have started out fresh and acquired saltiness with time. These techniques gave the Earth an age of roughly 100 million years, a date that was generally accepted by most nineteenth-century geologists.

Another approach was to calculate the time it took for the Earth to form out of the solar nebula and cool from a molten state to its present temperature. However, this method was soon discarded after the discovery of radioactive isotopes, which generated heat when they decayed into stable elements. This was responsible for maintaining the Earth's interior temperature since the very beginning. By calculating the half-lives of certain radioactive isotopes found on the Earth, moon and meteorites (Figure 54), scientists determined that the Earth is about 4.6 billion years old.

Figure 55 The upper horizontal plateau series is separated by an angular unconformity from the older, tilted Grand Canyon series. Photo by L. F. Noble, courtesy of USGS

ROCK CORRELATION

The seventeenth-century Danish physician and geologist Nicolaus Steno recognized that in a sequence of rock layers undeformed by folding or faulting each bed was formed after the one below it and before the one above it. This became known as the law of superposition. Steno also put forward the principle of original horizontality, which states that sedimentary rocks were originally laid in the ocean horizontally, and subsequent folding and faulting uplifted them out of the sea and inclined them at steep angles.

If angled rocks are overlain by horizontal ones, they represent a gap in time known as an angular unconformity (Figure 55). Furthermore, if a body of rocks cuts across the boundaries of other rock units, it has to be younger than those it intercepts. This is the principle of cross-cutting relationship, which implies that granitic intrusions are younger than the rocks they invade. A sequence of rocks placed in their proper sequence is called a stratigraphic cross section (Figure 56).

In order to develop a geologic time scale that is applicable over the entire world, rocks of one locality are matched or correlated with rocks of similar age in another locality. By correlating rocks from one place to another over a wide area, it is possible to obtain a comprehensive view of the geologic history of a region. A bed or a series of beds can thus be traced from one outcrop to another by recognizing certain distinctive features in the rocks.

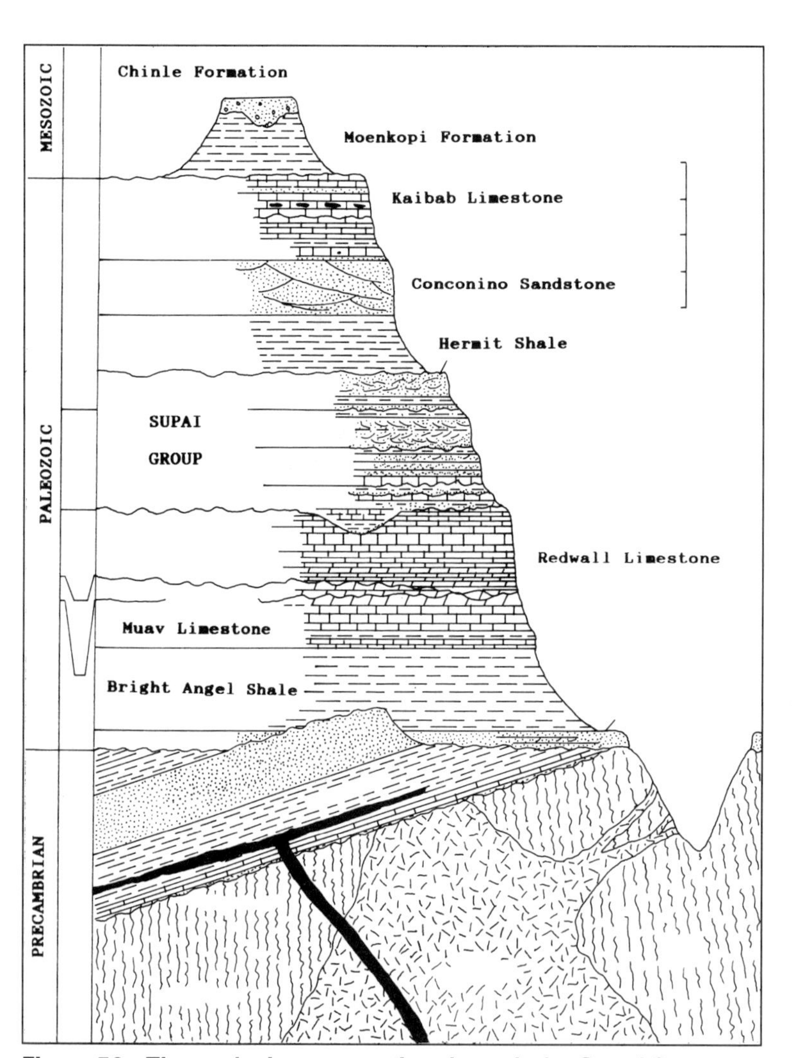

Figure 56 The geologic cross section through the Grand Canyon.

A problem arises, however, if there are two or more rock units at each locality that are identical, making it difficult to discern which bed matches with which. To further complicate matters, if there were faulting in the area, one block of a rock sequence might be downdropped in relation to the other or thrusted over another. A stratum folded over on itself contains rock units that are completely reversed, making matters even more confusing (Figure 57). Rocks that occur in repetitive sequences of sandstone, shale and limestone complicate correlation even further.

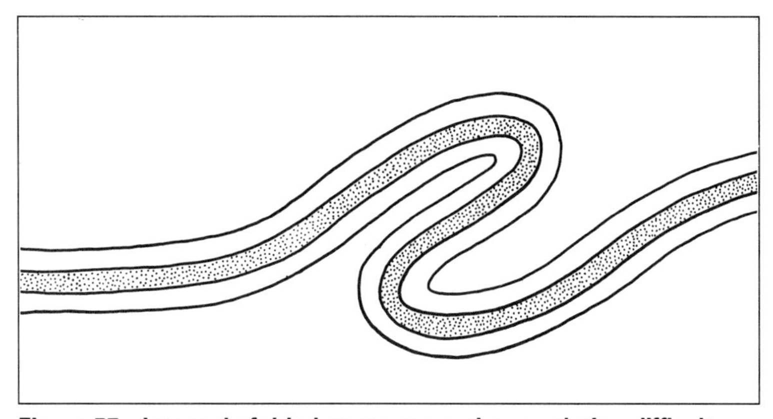

Figure 57 Intensely folded strata can make correlation difficult.

Although these methods might be sufficient to trace rock formations over relatively short distances, they are inadequate for matching rocks over long distances, such as from one continent to another. Therefore, in order to correlate between widely separated areas or between continents, fossils were used. Even though the existence of fossils has been known through the ages, it was not until the late eighteenth century that their significance as a geologic tool was discovered. The English engineer and geologist William Smith put forward the idea that two rock layers from different sites could be regarded as equivalent in age if they contained the same fossils. Thus sedimentary strata in widely separated areas could be identified by their distinctive fossil content.

DATING ROCKS

Before the advent of radiometric dating, there was no means of dating geologic events precisely. Therefore, relative dating techniques that relied on the fossil content of the rocks were developed and are still used. Absolute dating methods did not replace these techniques, only supplemented them. Since accurate absolute dates have been applied to periods of relative time, however, there have been some difficulties. There still remains some disagreement on the dates of certain events in geologic history.

The basic problem in attempting to assign absolute dates to units of relative time is that most radioactive isotopes are restricted to igneous rocks. Even if sedimentary rocks, which comprise most of the rocks on the

Earth's surface and hold all of the fossils, did contain a radioactive mineral, most rocks could not be dated accurately because the sediments are composed of grains derived from older rocks. Therefore, for sedimentary rocks to be dated, they must be related to igneous masses.

Volcanic ash deposited in a layer above or below a sediment bed could be dated radiometrically, as well as cross-cutting features such as a granitic dike, which is younger than the beds it crosses. The sedimentary strata would then be bracketed by dated materials, and its age could be determined fairly accurately. Today, the Earth's history is dated somewhat satisfactorily for most purposes, and it remains for further discoveries to improve the geologic time scale.

Of all the radioactive isotopes existing in nature, only a few have proved useful for dating ancient rocks (Table 7). The others are either very rare or have half-lives that are too short or too long. Rubidium 87 with a half-life of 47 billion years, uranium 238 with a half-life of 4.5 billion years and uranium 235 with a half-life of 0.7 billion years are useful for dating rocks that are tens of millions to billions of years old. The uranium isotopes are important for dating igneous and metamorphic rocks. Because both species of uranium occur together, they can also be used to cross-check each other.

TABLE 7 FREQUENTLY USED RADIOACTIVE ISOTOPES

Radioactive Parent	Half-life (years)	Daughter Product	Rocks and Minerals Commonly Dated
Uranium 238	4.5 billion	Lead 206	Zircon, uraninite, pitchblende
Uranium 235	713 million	Lead 207	Zircon, uraninite, pitchblende
Potassium 40	1.3 billion	Argon 40	Muscovite, biotite, hornblende, glauconite, sanidine, volcanic rock
Rubidium 87	47 billion	Strontium 87	Muscovite, biotite, lepidolite, microcline, glauconite, metamorphic rock
Carbon 14	5,730	Nitrogen 14	All plant and animal materials

Potassium 40 is more versatile for dating younger rocks. Although the half-life of potassium 40 is 1.3 billion years, recent analytical techniques make it possible to detect minute amounts of its stable daughter product argon 40 in rocks as young as 30,000 years old.

Dating sedimentary rocks presents a more difficult problem because their material was created by weathering processes. Fortunately, a micalike mineral called glauconite forms in the sedimentary environment and contains both potassium 40 and rubidium 87. As a result, the age of the sedimentary deposit can be established directly by determining the age of the glauconite. Unfortunately, metamorphism, no matter how slight, might reset the radiometric clock by moving the parent and daughter products elsewhere in the sample. Therefore, the radiometric measurement can only date the metamorphic event.

In order to date more recent events, scientists use a radioactive isotope of carbon called carbon 14, or radiocarbon. Carbon 14 is continuously created in the upper atmosphere by cosmic-ray bombardment of gases, which in turn release neutrons. The neutrons bombard nitrogen in the air, causing the nucleus to emit a proton, thus converting nitrogen to radioactive carbon 14. In chemical reactions, this isotope behaves just like natural carbon 12. It reacts with oxygen to form carbon dioxide, circulates in the atmosphere and is ab-

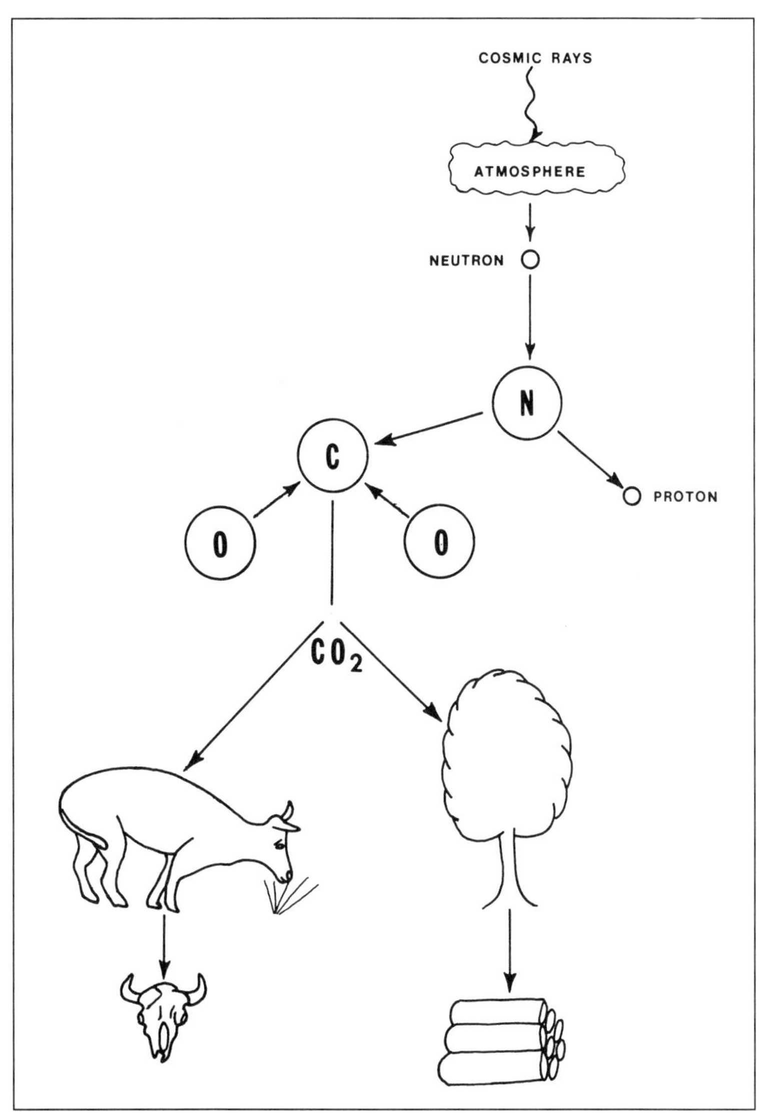

Figure 58 The carbon-14 cycle is important for dating previously living material.

sorbed directly or indirectly by living matter (Figure 58). As a result, all organisms, including humans, contain a small amount of carbon 14.

Carbon 14 decays at a steady rate with a half-life of 5,730 years. When an organism is alive, the decaying radiocarbon is continuously being replaced, and the proportions of carbon 14 and carbon 12 remain constant. However, when a plant or animal dies, it ceases to intake carbon, and the amount of carbon 14 gradually decreases as it decays to stable nitrogen 14. This results from the emission of a beta particle (free electron) from the carbon 14 nucleus, transmuting a neutron into a proton and restoring the nitrogen atom to its original state.

Radiocarbon dates are determined by chemical analysis, which compares the proportions of carbon 14 to carbon 12 in a sample. The development of improved analytical techniques has increased the usefulness of radiocarbon dating, and it can be used to date events that took place over 100,000 years ago. This makes it a valuable tool for dating events that occurred during the last ice age. Furthermore, paleontologists, anthropologists, archaeologists and historians now have a means of accurately dating events from humankind's distant past.

Figure 59 Sawatch sandstone resting on Precambrian granite on Ute Pass, El Paso County, Colorado. Photo by N. B. Darton, courtesy of USGS

GEOLOGIC FORMATIONS

Beginning at the very bottom of the geologic column are Precambrian granitic and metamorphic rocks (Figure 59). These are overlain by progressively younger sediments, igneous intrusives, metamorphics and volcanic extrusives. Deposited on top of the Precambrian basement rock are Proterozoic conglomerates, which are consolidated sand and gravel. Nearly 20,000 feet of Proterozoic sediments are found in the Uinta Range of Utah (Figure 60), which is the only major east-west-trending mountain range in North America. The Montana Proterozoic belt system contains sediments up to several miles thick.

The Proterozoic is also known for its terrestrial redbeds, so named because the sediment grains were cemented together with iron oxide, which stained the rocks red. In the western United States, there is a preponderance of red rocks exposed in the mountains and canyons. These sedimentary rocks were cemented by an iron oxide mineral called hematite, so named because of its blood red color. Redbeds are clear evidence that the Earth's atmosphere contained significant amounts of oxygen, which oxidized the iron in a way similar to rusting.

During the middle Paleozoic, as the continents rose higher and sea levels dropped lower, the inland seas disappeared and were replaced by great

Figure 60 Lodore Canyon, looking north toward Browns Park, Uinta Mountains, Summit County, Utah. Photo by W. R. Hansen, courtesy of USGS

Figure 61 Chugwater redbeds, Big Horn County near Shell, Wyoming. Photo by G. A. Fisher, courtesy of USGS

swamps. In these regions, thick coal deposits accumulated during the Carboniferous, which includes the Mississippian and Pennsylvanian periods in North America. The Carboniferous and Permian periods had the highest organic burial rates in Earth history. There was also a major ice age during the late Carboniferous.

The Permian witnessed the complete retreat of marine waters from the land,

Figure 62 Columbia River basalt, looking downstream from Palouse Falls, Franklin-Whitman Counties, Washington. Note the joint control of the stream. Photo by F. O. Jones, courtesy of USGS

an abundance of terrestrial redbeds and large deposits of gypsum and salt. In North America, terrestrial redbeds covered the Colorado Plateau (Figure 61) and a region from Nova Scotia to South Carolina. Redbeds were also common in Europe. The wide occurrence of red sediments might have resulted from the eruption of massive amounts of iron supplied by intense igneous activity the world over. Air trapped in ancient tree sap suggests that there was a greater abundance of atmospheric oxygen, which was responsible for oxidizing the iron into hematite.

Important reserves of phosphate used for fertilizers were laid down in the late Permian in Idaho and adjacent states. Huge sedimentary deposits of iron were also laid down, but they were not nearly as rich as those of the Precambrian. The ore-bearing rocks of the Clinton iron formation, the chief iron producer in the Appalachian region from Alabama to New York were deposited during this time.

An inland sea, called the Western Interior Cretaceous Sea, flowed into the west-central portions of North America, and accumulations of marine sediments eroded from the Cordilleran highlands to the west were deposited on the terrestrial redbeds of the Colorado Plateau, forming the Jurassic Morrison Formation, which is well known for its abundant dinosaur bones.

During the Cretaceous period, huge deposits of limestone and chalk were laid down in Europe and

Figure 63 Major John Wesley Powell. Courtesy of USGS

Asia. Seas invaded Asia, South America, Africa, Australia and the interior of North America. Into these vast bodies of water were deposited thick layers of sediment, which are presently exposed as impressive sandstone cliffs in the western United States.

At the beginning of the Tertiary period, volcanic activity was extensive and great outpourings of basalt covered Washington, Oregon and Idaho, creating the Columbia River Plateau (Figure 62). Massive floods of lava poured onto an area of about 200,000 square miles and in places reached 10,000 feet thick. In only a matter of days, volcanic eruptions spewed out batches of basalt as large as 1,200 cubic miles, forming lava lakes up to 450 miles across. Massive volcanism also occurred in other parts of the world, prompting the volcanic theory of dinosaur extinction, because such an environmental catastrophe would have made living conditions intolerable for many species.

Figures 64A and 64B Vibrosis vehicle and seismic instrument truck used for defining subsurface structures that might be responsible for trapping oil and gas. Courtesy of U.S. Department of Energy

GEOLOGIC MAPPING

Geologic maps display the distribution of rocks on the Earth's surface. They also indicate the relative ages of these formations and profile their position beneath the surface. Often, much of the information is compiled from just a few available exposures, which must be extrapolated over a large area. The first geologic maps were made by geologists in Britain, where one of their first practical uses was for the exploration of coal. In the western United States, early explorers were amazed by the magnificent rock exposures. Pioneer geologists like John Wesley Powell (Figure 63), who was the first to explore the Grand Canyon, made extensive geologic maps of the West, often sketching the formations on horseback as they rode through the region.

Modern geologic maps incorporate field observations and laboratory measurements, which are limited by rock exposures, accessibility and human resources. Regional geologic maps present rock composition, structure and geologic age, which are essential for constructing the geologic history of an area. Aided by geophysical data used for defining subsurface

Figure 65 Sedimentary dome near Sinclair, Wyoming. Photo by J. R. Balsley, courtesy of USGS

structures (Figures 64A and 64B), these reconstructions are important because formations of rock units and geologic structures influence the deposition of much of the mineral wealth of the world.

Remote sensing methods are primarily used for augmenting conventional techniques of compiling and interpreting geologic maps of large regions. With remote sensing techniques, it is possible to obtain certain structural and lithologic information much more efficiently than can be achieved on the ground. In well-exposed areas, geologic maps can be made from aircraft and satellite imagery, even if only limited field data are available, because many of the major structural and lithologic units are well displayed on the imagery.

Lineaments, which are long, linear trends in the Earth's surface, are one of the most obvious as well as most useful features in the imagery. Lineaments represent zones of weakness in the Earth's crust often formed as a result of faulting. Lineaments and texture, along with dip and strike of the strata, which are the degree and direction of formation slope, further aid in geologic mapping.

Other features frequently observed in the imagery include circular structures (Figure 65) created by domes, folds and intrusions of igneous bodies into the crust. Structural features such as folds, faults, dips and strikes of particular rock formations, along with lineaments, landform features, drainage patterns and other anomalies might suggest areas where oil traps exist.

Stream drainage patterns, which are influenced by topographic relief and rock type, give additional clues about the type of geologic structure. Furthermore, the color and texture of the structure carry information about the rock formations it comprises. With this information, it is possible to generate large-scale geologic maps of inaccessible areas. This has the potential of opening up entirely new areas for mineral and petroleum exploration, especially in an age when the need for resources is becoming highly critical.

6

FOSSIL BEDS

The history of the Earth is written in its rocks, and the history of life is told by fossils, the remains or traces of organisms preserved from the geologic past (Figure 66). However, the fossil record is not complete because the remaking of the surface by geologic forces has erased entire chapters of geologic time. Yet the study of fossils along with the radiometric dating of the rocks that contain them has enabled scientists to construct a reasonably good chronology of Earth history. The fossil record also provides valuable insights into the evolution of the Earth. Moreover, knowledge of the origination and extinction of species throughout the fossil record is important for building up an accurate account of the evolution of species through time (Table 8).

FAUNAL SUCCESSION

Although the existence of fossils has been known since the ancient Greeks first discovered them in the hillsides and pondered how they got there, the discovery of their significance as a geologic tool was not made until the late eighteenth century. The English civil engineer William Smith found that rock formations in the canals he built across Britain contained fossils

Figure 66 Early Paleozoic fossils on display at the Museum of Geology, South Dakota School of Mines at Rapid City.

that were different from those in beds above or below them. Smith also noticed that sedimentary strata in widely separated areas could be identified by their distinctive fossil content. These observations led him to propose the law of faunal succession, one of the most important and basic principles of historical geology.

Around the same time, the French geologists Georges Cuvier and Alexandre Brongniart found that certain fossils in the rocks around Paris were restricted to specific beds. The geologists arranged their fossils in chronological order and discovered that they varied in a systematic way according to their positions in the rock layers. Fossils in the higher rock layers more closely resembled modern forms of life than those further down the geologic column. Also, the fossils did not occur randomly but in a determinable order from simple to complex. Therefore, geologic time periods could be identified by their distinctive fossil content. This became the basis for establishing the geologic time scale (Table 9) and the beginning of modern geology.

When fossils are arranged according to their age, they do not present a random or haphazard picture but instead show progressive changes from

TABLE 8 EVOLUTION OF THE BIOSPHERE

Event	Billions of Years Ago	Percent Oxygen	Life Forms	Results
Full oxygen conditions	0.4	100	Fishes and land plants and animals	Approach present biological conditions
Appearance of shelled animals	0.6	10	Cambrian fauna	Burrowing habitat
Metazoans appear	0.7	7	Ediacarian fauna	First metazoan fossils and tracks
Eukaryotic cells appear	1.4	>1	Larger cells	Redbeds, multicellular organisms
Blue-green algae	2.0	1	Algal filaments	Oxygen metabolism
Algal precursors	2.8	<1	Stromatolites	Beginning photosynthesis
Origin of life	4.0	0	Light carbon	Evolution of biosphere

simple to complex life forms and reveal the advancement of species through time. Geologists were able to recognize geologic time periods based on groups of organisms that were especially plentiful and characteristic during a particular time. Within each period, there are many subdivisions determined by the occurrence of certain species, and this same succession is found on every major continent and is never out of order.

The branch of geology devoted to the study of ancient life based on fossils is called paleontology. Not all organisms become fossils, however, and plants and animals must be buried under stringent conditions to become fossilized. Given enough time, the remains of an organism are modified, often becoming petrified, literally turning to stone.

Fossils are important for correlating rock units over vast distances. Since certain species have lived only during specific times, their respective fossils can be used to place stratigraphic units in their proper sequence or relative time periods. These beds can then be traced over wide areas by comparing their fossil content. This provides a comprehensive geologic history over a broad region. It also establishes a relative time scale that can be applied to all parts of the world. (Absolute dates have since been placed on units of relative time to further improve our understanding of geologic history.)

RELATIVE TIME

One of the major problems encountered when exploring for fossils of early life is that the Earth's crust is constantly rearranging itself, and only a few

TABLE 9 GEOLOGIC TIME SCALE

Era	Period	Epoch	Age (millions of years)	First Life Forms
	Quaternary	Holocene	0.01	
		Pleistocene	2	Man
Cenozoic		Pliocene	7	Mastodons
		Miocene	26	Saber-tooth tigers
	Tertiary	Oligocene	37	
		Eocene	54	Whales
		Paleocene	65	Horses Alligators
	Cretaceous		135	
				Birds
Mesozoic	Jurassic		190	Mammals Dinosaurs
	Triassic		240	
	Permian		280	Reptiles
	Pennsylvanian		310	
	Carboniferous			Trees
Paleozoic	Mississippian		345	Amphibians Insects
	Devonian		400	Sharks
	Silurian		435	Land plants
	Ordovician		500	Fish
	Cambrian		570	Sea plants Shelled animals
			700	Invertebrates
Proterozoic Eon			2,500	Metazoans
			3,500	Earliest life
Archean Eon			4,000	Oldest rocks
			4,600	Meteorites

fossil-bearing formations have survived undisturbed over time. Therefore, the history of the Earth as told by its fossil record is incomplete because the remaking of the surface has erased parts of geologic history.

The existence of fossils has been known since ancient times, when Aristotle recognized that certain fossils were the remains of organisms, although he generally believed that fossils were the result of some celestial influence. This astrological account for fossils remained popular through the Middle Ages. It was not until the late Renaissance period and the rebirth of science that other explanations for the existence of fossils based on scientific principles were given. By the eighteenth century, most scientists began to accept fossils as the remains of organisms because they more closely resembled living things rather than merely inorganic substances.

Nineteenth-century geologists used fossils to define the boundaries of the geologic time scale. But because there was no means of actually dating the rocks that contained the fossils, the entire geologic record had to be delineated using units of relative time. This dating method only indicated which bed was older or younger in accordance to its fossil content. Therefore, relative dating only could place rocks in their proper sequence or order. It could not indicate how long ago an event took place, only that it followed one period and preceded another.

Geologists measure geologic time by tracing fossils through the rock strata and observing a greater change with deeper rocks as compared to those near the surface. Fossil-bearing strata can be followed horizontally over great distances because a particular fossil bed can be identified in another locality with respect to the beds above and below it. These are called marker beds, and they are important for identifying geologic formations. They were originally employed for the exploration of coal, one of the first practical uses of geology.

One of the most frustrating aspects of dealing with the geologic time scale are gaps in the fossil record, where entire chapters of Earth history have been erased. This might have

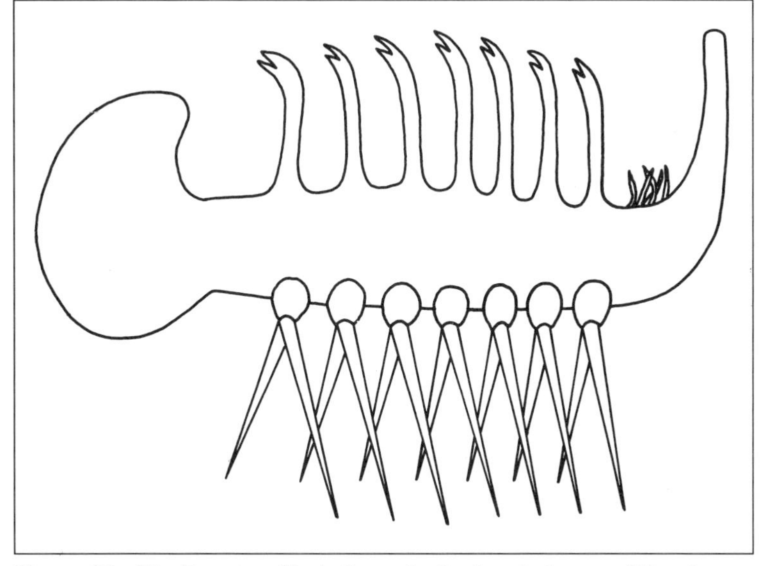

Figure 67 The Burgess Shale fauna hallucigenia is one of the strangest animals preserved in the fossil record. It walked on seven pairs of stilts and possessed seven tentacles, each with its own mouth.

TABLE 10
CLASSIFICATION OF SPECIES

Group	Characteristics	Geologic Age
Protozoans	Single-celled animals. Forams and radiolarians	Precambrian to recent
Poriferans	The sponges, about 3,000 living species	Proterozoic to recent
Coelenterates	Tissues composed of three layers of cells. About 10,000 living species. Jellyfish, hydra, coral	Cambrian to recent
Bryozoans	Moss animals. About 3,000 living species	Ordovician to recent
Brachiopods	Two asymmetrical shells. About 120 living species	Cambrian to recent
Mollusks	Straight, curled or two symmetrical shells. About 70,000 living species. Snails, clams, squids, ammonites	Cambrian to recent
Annelids	Segmented body with well-developed internal organs. About 7,000 living species. Worms and leaches	Cambrian to recent
Arthropods	Largest phylum of living species with over 1 million known. Insects, spiders, shrimp, lobsters, crabs, trilobites	Cambrian to recent
Echinoderms	Bottom dwellers with radial symmetry. About 5,000 living species. Starfish, sea cucumbers, sand dollars, crinoids	Cambrian to recent
Vertebrates	Spinal column and internal skeleton. About 70,000 living species. Fish, amphibians, reptiles, birds, mammals	Ordovician to recent

been due to periods of erosion or nondeposition of sedimentary strata that trap and preserve species as fossils. Gaps in the fossil record might also be attributed to insufficient intermediary species, or so-called missing links, which might have existed only in small populations. Small populations are less likely to leave a fossil record because the process of fossilization favors large populations of species, which are given the greatest recognition in the fossil record.

Figure 68 The coral-algal zone on a reef of the Bikini Atoll in the Pacific Marshall Islands. Photo by K. O. Emery, courtesy of USGS

FOSSIL CLASSIFICATION

Extinct organisms are classified by the same system used for classifying extant, or living, species. The classification scheme establishes a hierarchy of organisms, with each step up the evolutionary scale becoming more inclusive and encompassing a larger number of species. For example, a kingdom is the largest taxonomic classification unit and comprises all species of animals, plants or microorganisms.

A phylum, which is the next step below a kingdom, contains all organisms that share the same general body plan. There has been difficulty in placing some ancient organisms in existing phyla because their highly unusual body forms have no counterparts in today's living world. This is particularly true of the late Precambrian Ediacara fauna and the early Cambrian Burgess Shale fauna (Figure 67). Paleontologists thus had a difficult time classifying them into existing phyla, requiring the establishment of new phyla.

Only 10 phyla are necessary to classify the vast majority of animal life on Earth, both living and extinct (Table 10). The first phylum, Protozoa, includes the earliest and simplest life forms. Each succeeding phylum contains more complex species, concluding with the most complex phylum, Chordata, or the vertebrates, which includes us. The ordering of phyla in this manner recognizes the evolutionary advancement of species through time.

The first organisms to evolve were primitive bacteria and algae, which have existed for 3.5 billion years. They were followed by protozoans and sponges. The next evolutionary life forms were the coelenterates, including calcareous coral, which built impressive formations of limestone that trapped and fossilized other organisms to preserve them for all time. More recent corals are responsible for the construction of barrier reefs and atolls (Figure 68). The Great Barrier Reef off northern Australia is the largest structure built by living organisms.

Similar to corals but much smaller are the bryzoans, which are important marker fossils for correlating rock formations over long distances. The brachiopods, also called lamp shells because they look similar to old-style oil lamps, are among the most common fossils, with over 30,000 species

Figure 69 Molds and shells of mollusks on highly fossiliferous sandstone of the Glenns Ferry formation on Deadman Creek, Elmore County, Idaho. Photo by H. E. Malde, courtesy of USGS

cataloged in the fossil record. Perhaps these were the shells that baffled the ancient Greeks when they discovered them in the mountains.

The mollusks probably left the most impressive fossil record of all marine species (Figure 69). The phylum is so diverse it is often difficult to find common features among its members. The extinct ammonites were the most spectacular predators of the ancient seas and possessed spiral shells up to several feet in diameter (Figure 70).

The largest phylum of living organisms is the Arthropoda, and one of the first and best known of the ancient arthropods is the extinct trilobites, which were among the earliest organisms with hard shells (Figure 71). Many trilobites have bite scars, occurring mostly on the right-hand side of the fossil. Apparently, predators attacked mainly from the right side because the animal was more vulnerable there. Arthropods emerged from the ocean soon after plants began spreading over the continents. Centipedes and tiny, spiderlike arachnids are the oldest known land creatures, dating back about 415 million years.

The most complex land plants at that time grew less than an inch tall and would have looked like an outdoor carpet covering the landscape. However, primitive bacteria might have first colonized the continents as far back as 3 billion years ago. Before grasses and trees emerged, microbial soils might have made Earth more hospitable for life. The tiny organisms probably formed a dark, knobby soil, looking like lumpy mounds of brown sugar spread over the landscape.

Land plants rapidly evolved into more complex forms, and giant forests covered much of the continents during the Carboniferous period beginning around 350 million

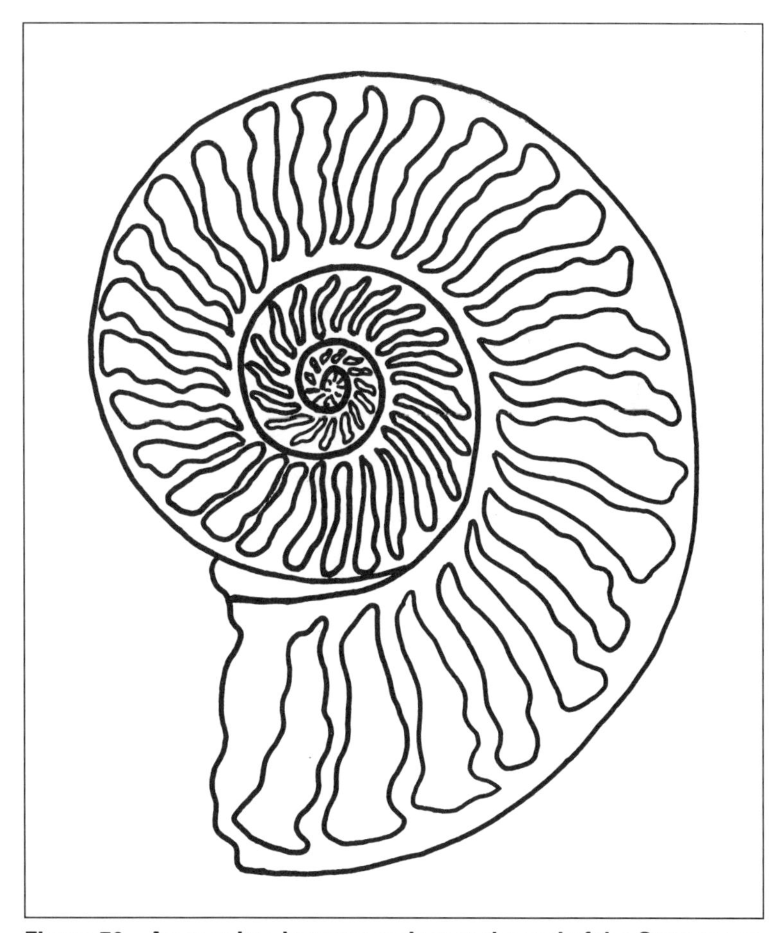

Figure 70 Ammonites became extinct at the end of the Cretaceous.

years ago, creating the world's great coal deposits. Flowering plants did not develop on a large scale until late in the Cretaceous. Because they were so successful, it is suggested that the plants might have had a hand in the demise of the dinosaurs by drawing down the level of atmospheric carbon dioxide, resulting in cooling conditions worldwide.

The echinoderms were possibly the strangest animals ever preserved in the fossil record. They are unique among species because of their radial symmetry and water vascular system used for feeding and locomotion. The great success of these creatures is illustrated by the fact that there are more classes of this animal both living and extinct than of any other phylum.

The higher animals are the vertebrates, which include fish, amphibians, reptiles, birds and mammals. The lobe-fin fish and lungfish were the first vertebrates to populate the land some 370 million years ago. These were the stem group from which all land vertebrates descended. The legacy of the vertebrates is well documented in the fossil record, and at no other time in geologic history were there so many varied and unusual creatures inhabiting the surface of the Earth.

During the Triassic period from 240 to 210 million years ago, several major groups of terrestrial vertebrates made their debut, including the ancestors of most modern reptiles, dinosaurs, mammals and possibly birds. The oldest bird fossil is of the 150-million-year-old *Archaeopteryx*, whose skeleton looked much like that of a small dinosaur.

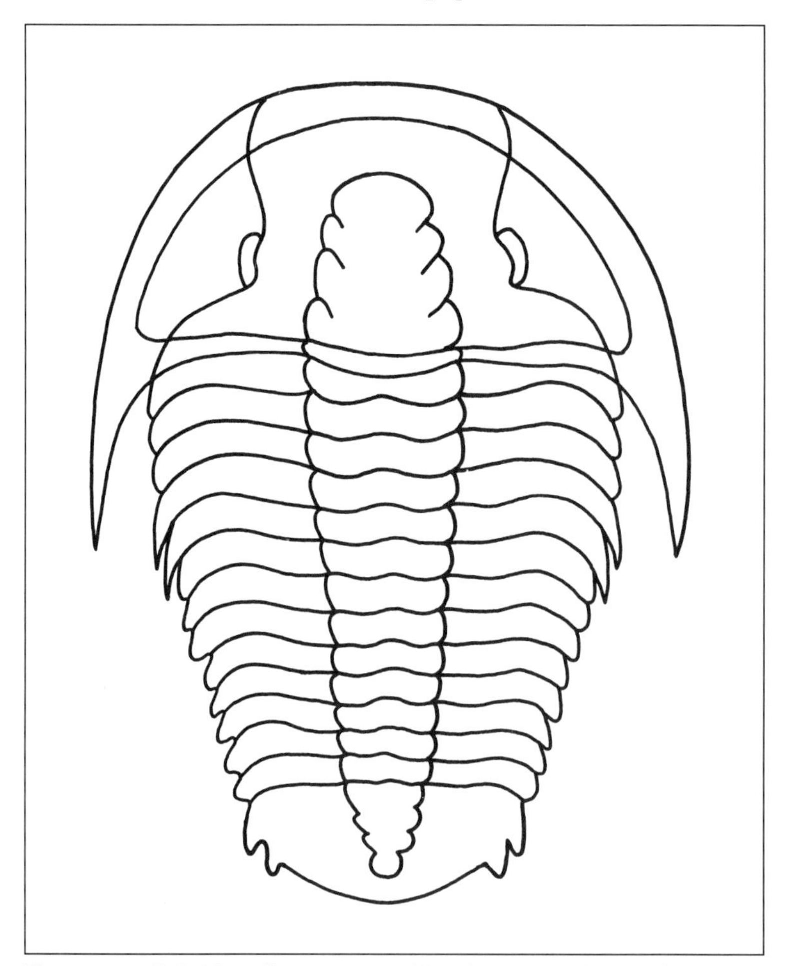

Figure 71 Trilobites first appeared in abundance in the Cambrian period and became extinct at the close of the Paleozoic era.

FOSSIL FORMATION

Geologic conditions that promote rapid burial with

minimal predation and decomposition aid in the preservation of dead organisms, and the eventual formation of fossils. Most fossils consist of ancient marine organisms because they were around for the longest time and therefore were the most plentiful. Thus, they stood a better chance of becoming fossilized. Also, because sedimentation occurs in the seas, they provide the best environment for the preservation of species. This makes marine organisms better candidates for burial than those living on land because continental erosion decreases the likelihood of burial and fossilization.

Figure 72 A paleontologist carefully chips away rock from a column of a dinosaur vertebrae at Dinosaur National Monument, Uintah County, Utah. Photo by W. R. Hansen, courtesy of National Park Service and USGS

Organisms with soft bodies, which lived before the arrival of shelled species at the beginning of the Cambrian period around 570 million years ago, had a difficult time entering the fossil record. When skeletons evolved, however, the number of organisms preserved in the fossil record jumped considerably. The introduction of hard parts has been called the greatest discontinuity in the history of life and signaled a major change in Earth history by accelerating the pace of development of new organisms. At this time, nearly all forms of life known today appeared in the fossil record, which is why the period is often called the "Cambrian explosion."

When oceanic calcium levels increased due to continental erosion, skeletons were developed as receptacles for early soft-bodied creatures, which had to dispose of the excess mineral from their tissues because of its toxicity. As concentrations of calcium in the ocean further increased, animal skeletons became more diverse and elaborate. Levels of atmospheric oxygen also appeared to rise in concert with the skeletal revolution. Skeletons also evolved as a response to an incoming wave of predators. Paradoxically, most of these predators were soft-bodied and therefore not preserved.

It should not be surprising then that fossils are dominated by organisms with hard skeletal remains (Figure 72). As a result, shells, bones, teeth and wood prevail in the record of past life. Unfortunately, because preservation is contingent on these demanding conditions, the fossil record is well represented by organisms with hard body parts but is lacking in organisms possessing soft bodies. This gives the fossil record a somewhat lopsided view of previous life on Earth.

Only a small portion of all life has been preserved as fossils. The majority of plant and animal remains is destroyed through predation and decay. Although it seems that becoming a fossil is not too difficult for some organisms, for others it is nearly impossible. Furthermore, only a small percentage of all fossils are exposed on the surface. Most of these are destroyed by weathering processes, which makes for an incomplete fossil record, leaving many species poorly represented or missing entirely.

In order for organisms to be preserved, they must possess hard body parts such as shells or bones, which is why these objects predominate in the fossil record. Soft fleshy parts are quickly destroyed by predation or bacterial decay. Even hard parts left on the surface for any duration are destroyed. Therefore, organisms have to be buried rapidly in order to escape destruction by the elements and to find protection against weathering and erosion.

Figure 73 Carbonized plant remains of a middle Pennsylvanian fern of the Kanawha series, West Virginia. Photo by E. B. Hardin, courtesy of USGS

Delicate organisms such as insects are extremely difficult to preserve, and consequently they are quite rare in the fossil record. Not only do they require protection from decay, but they must not be subjected to pressures that could crush them. Leaves and other delicate organisms are best preserved by carbonization, whereby a thin carbon film between sediment layers outlines the organism (Figure 73). Impressions formed in fine sediment layers are another means of preserving the general outlines of organisms. Some species such as the late Precambrian Ediacara fauna of Australia and the early Paleozoic Burgess Shale fauna of western Canada are known only by their impressions, which are among the most spectacular in the world.

The bones of extinct animals are much rarer than their footprints, and many species are known only by their tracks and trails. Even clear outlines of claws, the shape of the footpad and the pattern of scales are easily recognizable. Much information about an animal's lifestyle can be determined by analyzing its footprints, including its mode of locomotion, its gait and speed, and social habits—whether it was solitary or traveled in herds.

Dinosaur tracks are among the most impressive of all fossil footprints (Figure 74), mostly because of the animal's great weight, which left deep indentions in the ground. Their footprints are found in relative abundance in terrestrial sediments of the Mesozoic age throughout the world. Much information can be garnered by studying dinosaur tracks,

Figure 74 Dinosaur trackway in Tarapaca Province, Chile. Photo by R. J. Dingman, courtesy of USGS

Figure 75 Rhythmically interbedded limestone and shaly limestone of the Nasorak Formation, Lisburne district, northern Alaska. Photo by R. H. Campbell, courtesy of USGS

which suggests that some species were highly social. Large carnivores like *Tyrannosaurus rex* were swift, agile dinosaurs that could run up to 45 miles per hour, which is determined by measuring their fossil tracks. Contrary to popular belief, the dinosaurs were a successful species, existing on earth for 160 million years and evolving into about 1,000 distinct genera.

FOSSILIFEROUS ROCKS

Outcrops of marine sediments are generally the best sites for finding fossils. In most parts of the world, the central portions of the continents were invaded by seas during various times in the geologic past, allowing thick deposits of marine sediments to accumulate. Even the presently high continental interiors were once invaded by inland seas. When the seas departed and the land was uplifted, erosion exposed many of these marine sediments.

Limestones are among the best-suited rocks for fossil preservation mainly because of the nature of their sedimentation, often involving shells and skeletons of dead marine life that were buried and fused into solid rock. Most limestones originated in marine environments and some were deposited in lakes. Limestones constitute approximately 10 percent of all exposed sedimentary rocks, and many limestones form well-exposed outcrops (Figure 75).

FOSSIL BEDS

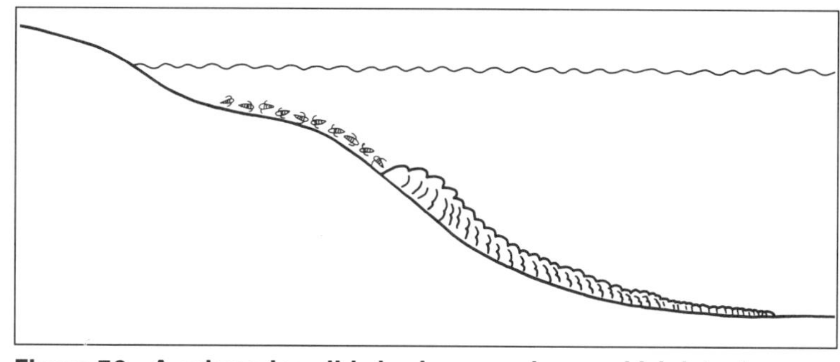

Figure 76 A submarine slide buries organisms, which later become fossilized.

Most limestones contain whole or partial fossils, depending on whether they were deposited in quiet or agitated waters. Tiny, spherical grains called oolites are characteristic of agitated water, whereas lithified layers of limey mud called micrite are characteristic of calm waters. In quiet waters, undisturbed by waves and currents, whole organisms with hard body parts are buried in the calcium carbonate sediments and are later lithified into limestone.

Shale and mudstones commonly contain fossils. They are the most abundant sedimentary rocks because these sediments are the main weathering products of feldspars, the most abundant minerals. Because clay particles sink slowly in the sea, they normally settle out in calm, deep waters far from shore. Compaction squeezes out the water between sediment grains, and the clay is lithified into shale. Organisms caught in the clay are compressed into thin carbonized remains or impressions.

Figure 77 A quarry in Bangor limestone near Russellville, Alabama. Photo by E. F. Burchard, courtesy of USGS

The deep, calm bottom waters are often stagnant and oxygen-poor. Periodic slumping from a high bank often results in a flow of mud into the deeper waters. Organisms living on or in the shallow muddy bottom are caught in the slide and buried in the mud when the slide comes to a halt in deeper water (Figure 76). Because scavengers cannot survive in these waters, the remains of the organisms are favored for preservation. As the mud gradually compacts and becomes hard rock, the buried carcasses are flattened into dark carbonized films. Fossilization in this manner can also preserve soft parts, whereas they are not as well preserved in limestone formations.

Sandstones contain well-preserved terrestrial fossils and imprints and faithfully record the passage of animals by their fossil footprints. These are best made with deep impressions in moist sand that are filled with finer

Figure 78 A French thrust in a roadcut, which places the lower beds of the Castle Reef Dolomite onto dark gray shale on the west side of French Gulch near the Sun River, Lewis and Clark County, Montana.
Photo by R. R. Mudge, courtesy of USGS

sediments such as windblown sand and later buried and lithified into sandstone. Subsequent erosion exposes the layers of sandstone, and the softer material that originally filled the depression weathers out, often exposing a clear set of footprints.

FINDING FOSSILS

Limestone is used in the manufacture of portland cement for building and road construction across the nation, where numerous limestone quarries exist (Figure 77). Abandoned limestone quarries are good sites for collecting fossils because many layers of limestone are penetrated. The rocks have been conveniently broken up, exposing fresh surfaces on which fossils are clearly seen. Abandoned coal pits are also numerous throughout the land and are excellent locations for collecting fossils of plants and animals that were buried in the great ancient coal swamps.

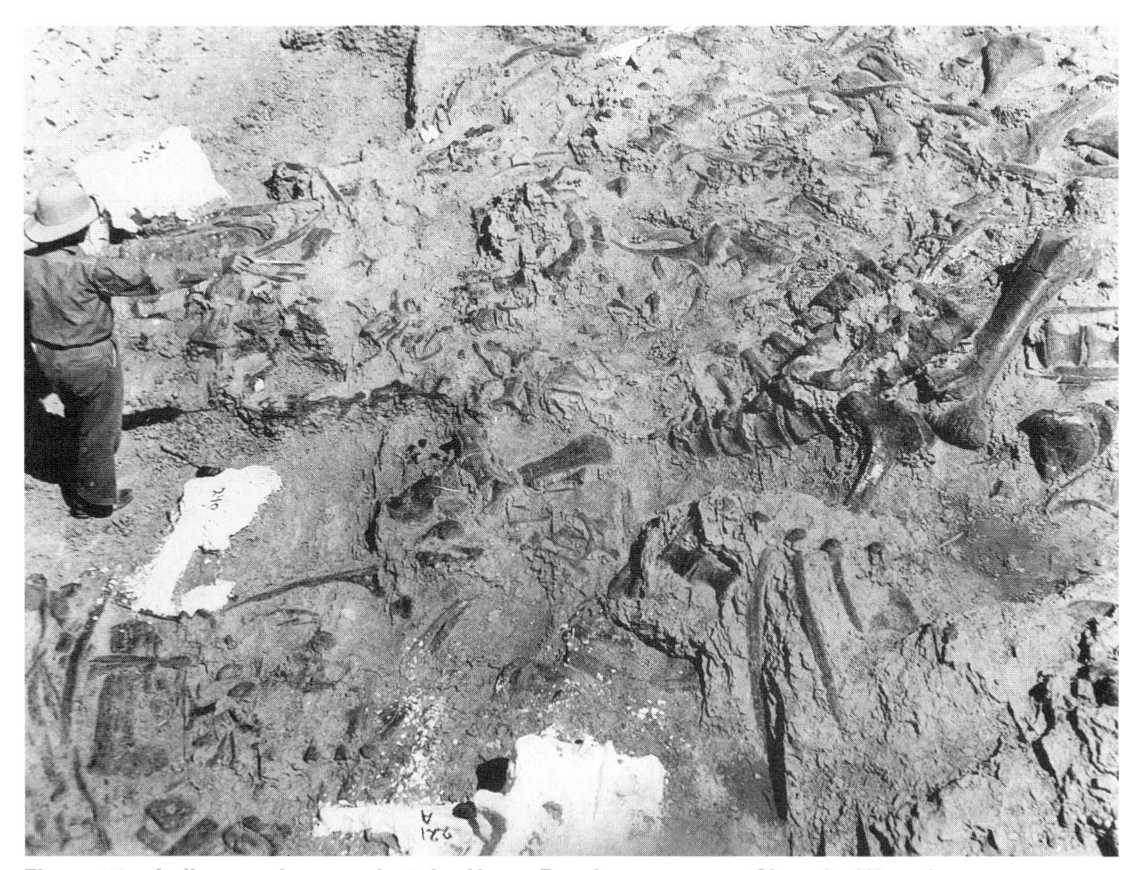

Figure 79 A dinosaur boneyard at the Howe Ranch quarry near Cloverly, Wyoming. Photo by G. E. Lewis, courtesy of USGS

Road cuts provide perhaps the most accessible rock exposures (Figure 78). When a road is constructed through hilly or mountainous terrain, often huge quantities of rock are blasted away, exposing hidden formations. Much of the geology across the country has been mapped using road cuts, railway cuts and tunnels, and sometimes these are the only good rock exposures. If a road is cut through limestones or shales, there is a good chance that fossils might be present. Stream channels are less accessible but have equally good rock exposures, especially those that cut deep ravines.

Perhaps the easiest fossil collecting is conducted in areas where specimens have been weathered out of the rock and lie in loose sediment or rocky debris at the base of an exposure, called scree or float. Often the limestone encasing a fossil erodes more easily, leaving whole specimens scattered on the ground. Broken-up rock in abandoned limestone quarries might also provide excellent fossils.

Major paleontologic digs give amateur fossil hunters an opportunity to work alongside professionals in excavating dinosaur bones (Figure 79). Rabbit Valley west of Grand Junction, Colorado, is a dinosaur quarry with easy access, lying just off the interstate highway, where for more than 100 years fossil collectors have been striking it rich. The quarry is responsible for the discovery of one of the largest dinosaur species, called *Apathosaurus*. This gargantuan creature fully deserves the title "terrible lizard."

7

FOLDING AND FAULTING

Tectonics, from the Greek *tekton* meaning to "build," are responsible for the Earth's active geology, including tall mountains, long rift valleys and deep-ocean trenches. Tectonics create the most impressive geologic features on the planet, using the forces of uplift combined with erosion. The crust can also be sliced by faults, which relieve pressures building up from plate motions. Slippage along major fault systems is often accompanied by earthquakes that can be very destructive as they wrench the landscape apart. It is therefore not surprising that many places on Earth are indeed on shaky ground.

TECTONISM

The Earth's outer shell is fashioned out of 7 major and several minor movable plates that account for all tectonic activity taking place on the surface of the planet. The plates are composed of the lithosphere, which is the rigid outer layer of the mantle, along with the overlying continental or oceanic crust. Because continental crust is made of light materials, it remains on the surface where it continues to grow, collecting more materials. The oceanic crust, on the other hand, because it is denser, subducts deep into the mantle at ocean trenches where it remelts in a continuous cycle.

The plate boundaries are the spreading ridges where new oceanic crust is created, the deep-sea trenches where old oceanic crust is subducted into the mantle and destroyed, and the transform faults where plates slide past each other. The plates carry the continents along with them as they ride on the semimolten rocks of the asthenosphere, the layer beneath the lithosphere. When two plates collide, they create mountain ranges on the continents and volcanic island arcs on the ocean floor. When an oceanic plate subducts beneath a continental plate, it forms sinuous mountain chains, like the Andes of South America, and volcanic mountain ranges, like the Cascades of the Pacific Northwest (Figure 80). The breakup of a plate creates new continents and oceans. The process of rifting and patching of continents has been going on for at least 2.7 billion years.

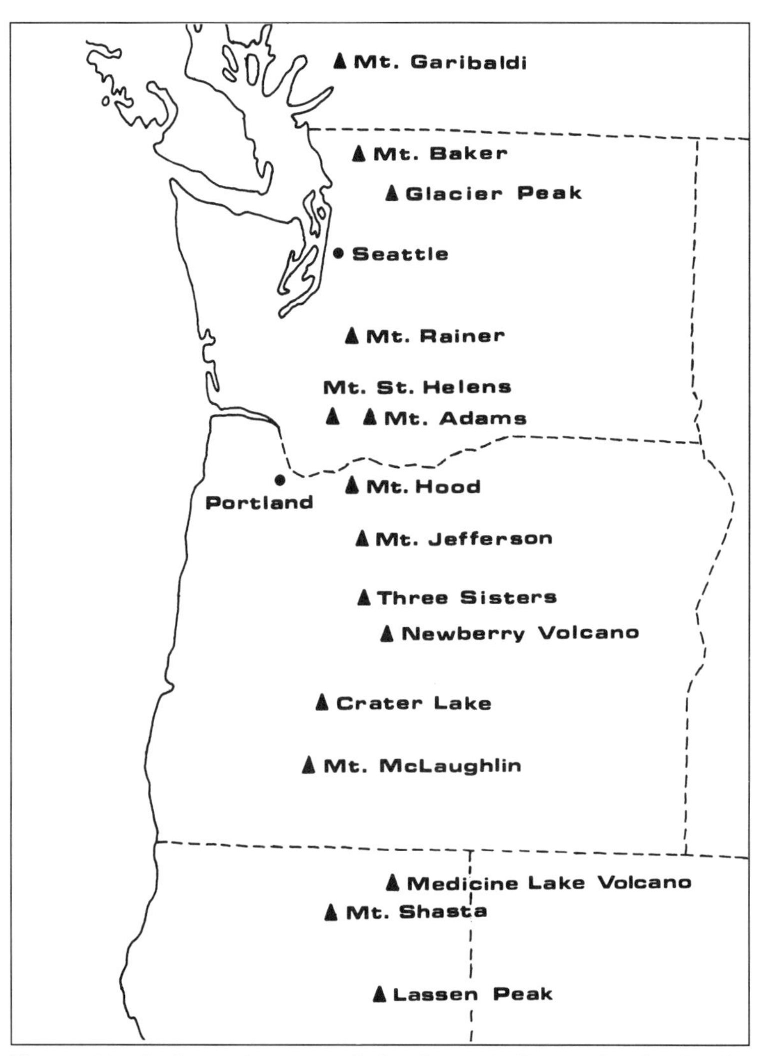

Figure 80 Active volcanoes of the Cascade Range in the Pacific Northwest.

An impressive submarine mountain range known as the Mid-Atlantic Ridge runs along the middle of the Atlantic Ocean, surpassing in scale the Alps and Himalayas combined. It is part of a global spreading-ridge system that stretches over 40,000 miles along the ocean floor like the stitching on a baseball. A deep trough is carved down the middle of the ridge and resembles a giant crack in the Earth's crust. It is the longest and deepest canyon on Earth.

The Mid-Atlantic Ridge is also the center of intense seismic and volcanic activity and the focus of high

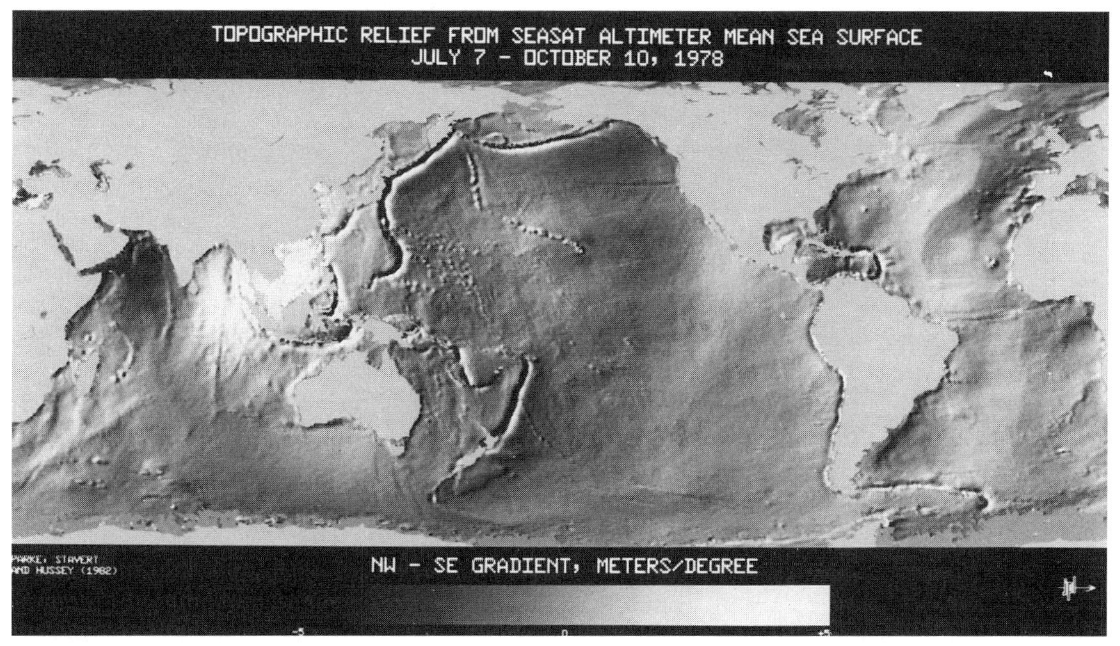

Figure 81A Topographic relief map of the ocean surface, showing features on the ocean floor including midocean ridges and trenches. Courtesy of NASA

heat flow from the Earth's interior. Molten magma originating from the mantle rises through the lithosphere and erupts on the ocean floor, adding new oceanic crust to both sides of the ridge crest. Meanwhile, the upwelling magma pushes apart the two lithospheric plates, upon which ride the continents that surround the Atlantic Ocean.

As the Atlantic Basin widens, it separates the continents surrounding the Atlantic Ocean at a rate of about an inch per year, or about the same rate as a fingernail grows. The spreading of the ocean floor in the Atlantic compresses the seafloor in the Pacific to make more room. The Pacific Basin is ringed by subduction zones (Figures 81A and

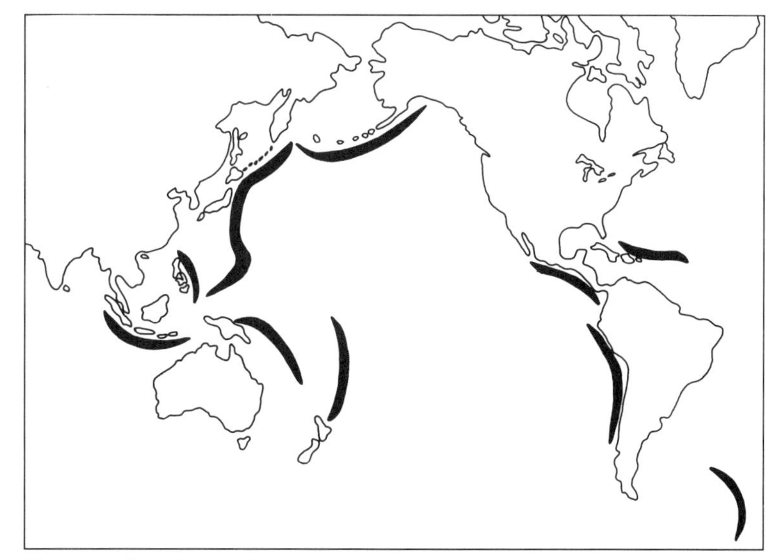

Figure 81B The major trenches of the world.

81B), which swallow up oceanic plates. These are responsible for most of the geologic activity that fringes the Pacific Ocean.

Three plates that border the Pacific—the Nazca, Antarctic and South American—come together in an unusual triple junction. The first two plates are spreading apart along a boundary called the Chile Ridge. The Nazca plate is being subducted beneath the South American plate at the Peru-Chile Trench and will eventually disappear. Spreading ridges in the Pacific are much more active than those in the Atlantic. Rapid spreading centers give rise to much lower relief on the ocean floor than slower ones because lava is unable to form tall heaps.

If placed end to end, all subduction zones would stretch clear around the world. The subduction of the lithosphere into the mantle plays a pivotal role in global tectonics and accounts for many of the geologic processes that shape the surface of the planet. The seaward boundaries of the subduction zones are marked by the deepest trenches in the world, which are found at the edges of continents or along volcanic island arcs. Major mountain ranges and most volcanoes and earthquakes are associated with the subduction of lithospheric plates. When plates thicken, they grow too heavy to remain on the surface and sink into the mantle, forming a long line of subduction represented by a deep trench. The sinking of a plate is also the main driving force behind continental drift, and pull at subduction zones is favored over push at spreading ridges to move the continents around.

The oceanic crust is composed of basalts that originated at spreading ridges and sediments washed off the continents. When the crust along with the underlying lithosphere is subducted into the mantle, it melts and the molten rock rises toward the surface. When the magma reaches the base of the continental crust, it becomes the source material for volcanic and magmatic activity. The magma also erupts on the ocean floor, forming long chains of volcanic islands. In this manner, plate tectonics is continuously changing and rearranging the face of the planet.

MOUNTAIN BUILDING

Mountains are areas of high relief, rising abruptly above the surrounding terrain. Folded mountain belts created by the collision of continental plates constitute a massive deformation of the rocks that form the core of the range. Most mountains are built by plate motions, which shove the crust of one plate onto another plate. Many peaks form like the Wind River Mountains of Wyoming, beneath which a gently sloping fault indicates that horizontal squeezing of the continents and not vertical lifting is responsible.

Mountains are also created when a deep root of light crustal rock literally floats the mountain like an iceberg. Additional buoyancy might be provided when the underlying lithosphere drips away from the crust and is replaced by hot rock from the mantle, further pushing the mountain range upward. Globs of relatively cold rock dropping hundreds of miles into the mantle appear to precede this type of mountain building. A good example is the 2.5-mile-high southern Sierra Nevada Range, which has risen about 7,000 feet over the last 10 million years, yet no plates have converged near the region for more than 70 million years.

Thousands of feet of sediments are deposited along the seaward margin of a continental plate in deep-ocean trenches, and the increased weight presses downward on the oceanic crust. As continental and oceanic plates merge, the heavier oceanic plate is subducted under or overridden by the lighter continental plate, forcing it further downward. The sedimentary layers of both plates are compressed, resulting in a swelling of the leading edge of the continental crust. This process forms a mountain belt similar to the Andes of South America (Figure 82). The topmost layers are scraped off the descending oceanic crust and plastered against the swollen edge of the continental crust. In the deepest part of the continental crust, where temperatures and pressures are extremely high, rocks are partially melted and metamorphosed. Pockets of magma also provide new source material for volcanoes and other igneous activity.

Magma extruding onto the Earth's surface builds volcanic structures such as broad plateaus and mountains. The volcanoes of the Cascade Range were created by the subduction of the Juan de Fuca plate along the Cascadia subduction zone beneath the northwestern United States. As the plate melts while diving into the mantle, it feeds molten rock to magma chambers underlying the volcanoes. Besides supplying magma for this string of hungry volcanoes, the subducting plate also has the potential of generating very strong earthquakes in the region.

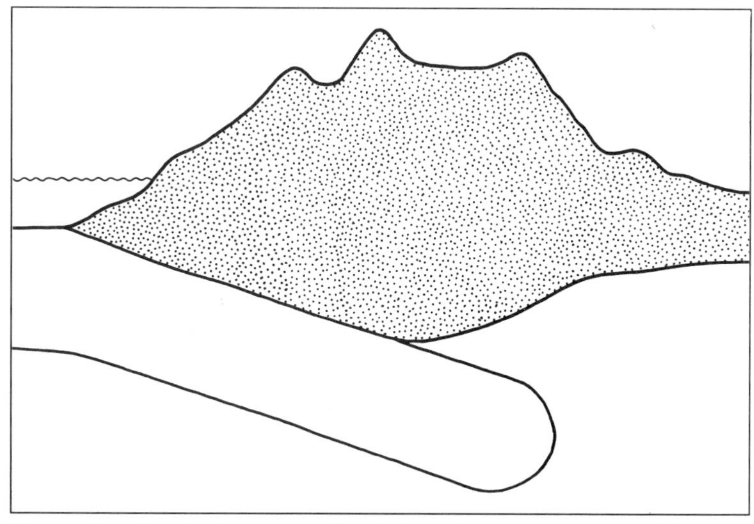

Figure 82 The Andes were raised due to increased crustal thickness as the Nazca plate plunges under South America.

The continental roots underlying mountain ranges can extend downward 100 miles or more into the upper mantle. Because of collisions arising from plate tectonics, continents have stabilized part of the mobile mantle rock beneath them. The drifting continents are thus able to carry along with them thick layers of chemically distinct mantle rock. The process that forms deep roots operates by squeezing a plate into a thicker one by continental collision. Nowhere is this process more intense than the collision between the Indian plate and the Eurasian plate; the latter has shrunk about 1,000 miles and in the process raised the great Himalayas, the world's tallest mountains, and the Tibetan Plateau, the largest tableland in the world.

FOLDED STRATA

Prior to the introduction of the plate tectonics theory, the formation of mountain ranges was still very much a mystery. It was generally thought that mountains formed early in the Earth's history when the molten crust solidified and shriveled up like a baked apple. After making more extensive studies of mountain ranges, however, geologists were forced to conclude that the folding of rock layers was much too intense (Figure 83), requiring considerably more rapid cooling and contraction than was possible. Moreover, if mountains were formed in this manner, they would have been scattered evenly throughout the world instead of concentrating in a few chains.

Most mountains occur in ranges; although a few isolated peaks do exist, they are rare. Mountains have complex internal structures formed by folding, faulting, volcanic activity, igneous intrusion and metamorphism. Mountain building, which provides the forces necessary for folding and faulting rocks at shallow depths, also provides the stresses that strongly distorts rocks at greater depths.

Wind and water gradually remove almost all signs of what were once splendid mountain ranges. But erosion does not erase everything, and often the roots of ancient mountains survive beneath what looks like unimpressive landscape. Buried are huge faults and folds that run through the basement rock, which indicate that long ago tectonic forces squeezed the crust, causing mountains to grow. Similar folds and faults form the roots of more modern ranges like the Appalachians and Rockies.

Figure 83 Intensely folded Cambrian carbonate rocks of Scapegoat Mountain, Lewis and Clark County, Montana. Photo by M. R. Mudge, courtesy of USGS

Vermont still preserves the roots of ancient mountains shoved upward some 400 million years ago, when the proto–North American and African continents collided. It is unknown how long the squeezing, heating and alteration of rock went on within the new mountains. However, according to analysis of radioactive elements, the process of mountain folding appears to take place over only a few million years.

When continents collide, they crumple the crust and force up mountain ranges at the point of impact. The sutures joining the landmasses are preserved as eroded cores of ancient mountains called orogens, from the Greek *oros*, meaning "mountain." Many of today's folded mountain ranges were uplifted by Paleozoic continental collisions that raised huge masses of rocks into several mountain belts throughout the world (Figure 84).

The Appalachians (Figure 85) were formed when North America and Africa slammed into each other during the late Paleozoic. The southern Appalachians are underlain by over 10 miles of sedimentary

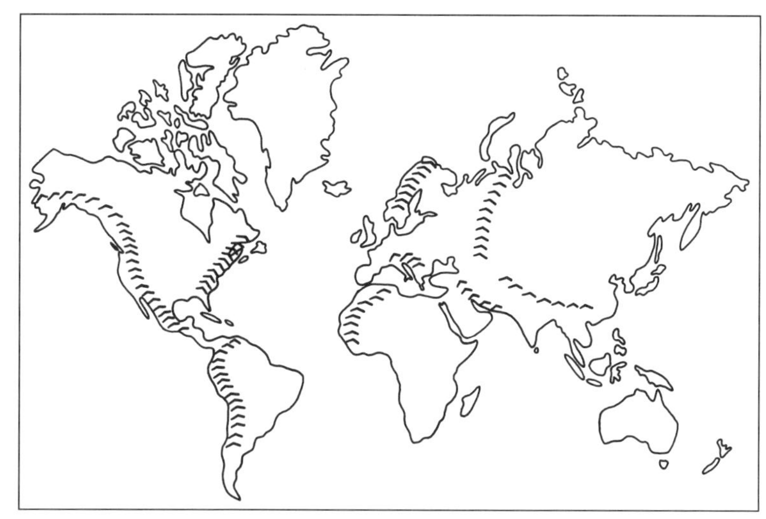

Figure 84 Major mountain ranges resulting from continental collisions.

Figure 85 The Appalachian Mountains were created by the collision of North America and Africa from the late Devonian to the late Permian periods.

and metamorphic rocks that are essentially undeformed, whereas the surface rocks were highly deformed by the collision due to thrust faulting. Caught between the colliding continents was the proto–Atlantic Ocean, also called the Iapetus, which was squeezed completely dry.

About 50 million years ago, the Tethys Sea between Africa and Eurasia began to narrow as the two continental plates collided, and around 20 million years ago it was closed off entirely. Like a rug thrown across a polished floor, the crust folded over into giant pleats. Thick sediments accumulating in the sea for tens of millions of years were compressed into long belts of mountain ranges on the northern and southern continental landmasses. The entire crusts of both continental plates buckled upward, forming the central portions of the ranges. The Alps formed in much the same manner as the Himalayas when the Italian prong of the African plate plunged into the Eurasian plate when the two continents came together.

Additional compression and deformation might take place farther inland beyond the line of collision, creating a high plateau with surface volcanoes, similar to the wide plateau of Tibet, which lies at an average elevation of over 3 miles above sea level. The strain of raising the world's highest mountain range by the collision of the Indian plate with the Eurasia plate has resulted in deformation and earthquakes all along the plate. India is still plowing into Asia at a rate of about 2 inches a year. As resisting forces continue to build up, plate convergence will eventually stop and the mountains will cease growing and begin to lose the battle with erosion.

FAULT TYPES

The mechanism behind earthquakes was poorly understood until after the great 1906 San Francisco earthquake. For hundreds of miles along the San Andreas Fault, fences and roads crossing the fault were displaced by as much as 21 feet. The San Andreas Fault is a 650-mile-long, 20-mile-deep fracture zone that runs northward from the Mexican border through southern California and represents a zone of separation between the North American and Pacific plates.

During the San Francisco earthquake, the Pacific plate suddenly slid as much as 21 feet northward past the North American plate. During the 50 years prior to the earthquake, land surveys indicated displacements as much as 10 feet along the fault. Tectonic forces were slowly deforming the crustal rocks on both sides of the fault, causing huge displacements. Meanwhile, the rocks were bending and storing up elastic energy. Eventually, the forces holding the rocks together were overcome, and slippage occurred at the weakest point.

The San Andreas Fault is perhaps the best-studied fault system in the world. It covers much of California, separating southwestern California

Figure 86 Collapsed building in the Marine District caused by the October 17, 1989, Loma Prieta earthquake, San Francisco County, California. Photo by G. Plafker, courtesy of USGS

from the rest of the North American continent. The segment of California west of the San Andreas Fault along with the lithospheric plate on which it rides is slipping past the continental plate in a northwesterly direction at a rate of about 2 inches per year.

The relative motion of the two plates is called right-lateral or dextral movement because an observer on either side of the fault would notice the other block moving to the right. If the two plates slid past each other smoothly, Californians would not worry so much about earthquakes. Unfortunately, especially in the southern end of the fault and in an area known as the Big Bend in the northern part of the fault, the plates tend to snag. When they attempt to tear themselves free, earthquakes rumble across the landscape. Both the 1906 San Francisco earthquake and the 1989 Loma Prieta earthquake (Figure 86) took place on a segment of the San Andreas Fault that runs through the Santa Cruz Mountains. Several milder aftershocks are expected within a year after such major earthquakes.

If California were reconstructed as it was 30 million years ago, when the northern extension of the East Pacific Rise first came to intercept the North

American continent, the segment west of the San Andreas Fault would have been south of the present Mexican border. If this motion continues for another 30 million years, southwest California could end up south of the present Canadian border. No catastrophic earthquake could ever send southern California crashing into the sea, however. Instead, it will continue on its slow journey northward, and in 50 million years the plate on which it rides will disappear down the Aleutian Trench while the crust is plastered against Alaska.

Subsidiary faults along the San Andreas (Figure 87) include numerous parallel faults such as the Hayward that runs through suburban San Francisco, the Newport-Inglewood and numerous transverse faults. The Garlock Fault is a major east-trending fault. Movement along this fault is left-lateral, or sinistral, and, combined with the right-lateral movement of the San Andreas Fault, is causing the Mojave Desert to the south to move eastward with respect to the rest of California. The faults of the Mojave and adjacent Death Valley absorb about 10 percent of the total slippage between the Pacific and North American plates. The complex crustal movements associated with these faults are responsible for most of the tectonic and geologic features of California such as the Sierra Nevada and the Coastal Ranges. Moreover, most of the earthquakes that plague California are produced by these faults.

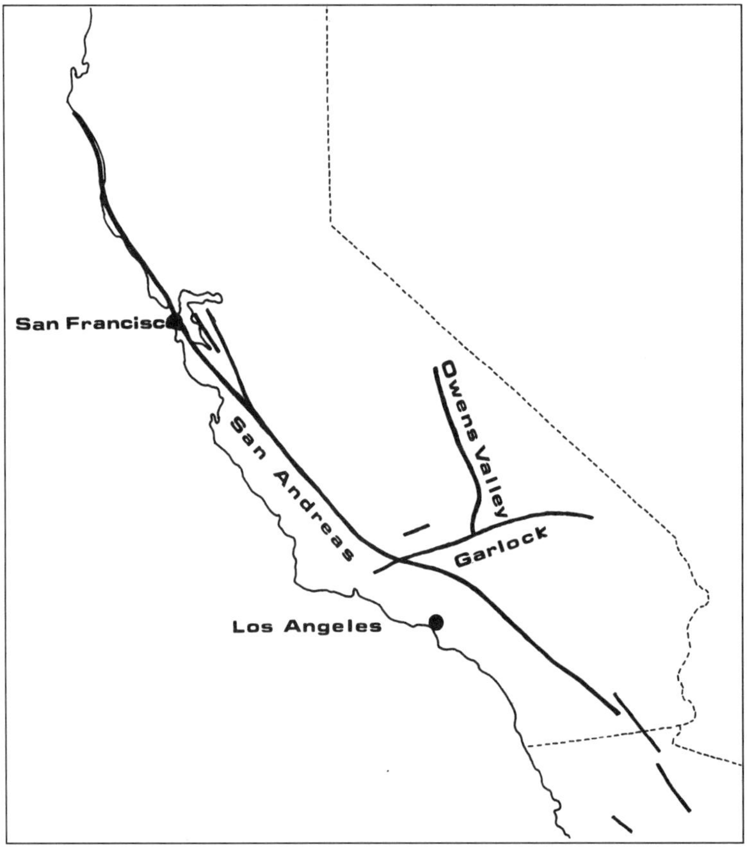

Figure 87 The San Andreas Fault and associated faults in California.

Not all movement along faults is horizontal. Vertical displacements, with one side of the fault positioned higher than the other side, are also common. Faults are classified by the relationship of the rocks on one side of the fault plane with respect to those on the other side (Figure 88). If the crust is pulled apart, one side of the fault will slide downward past the other side along a plane that is often steeply inclined.

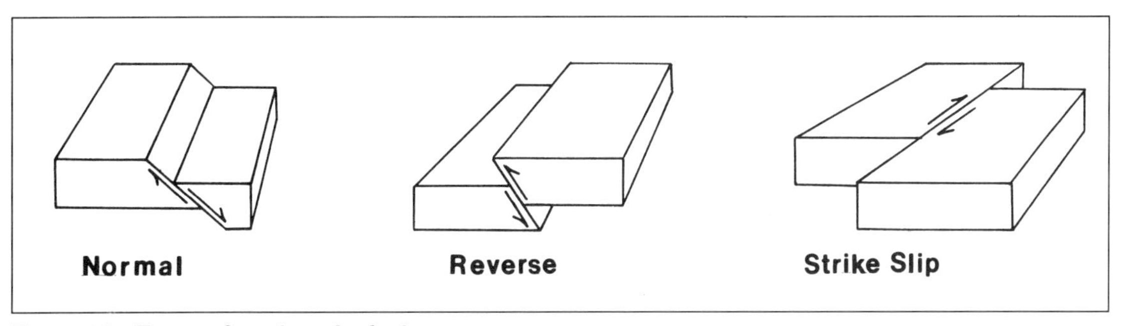

Figure 88 Types of earthquake faults.

This is known as a gravity or normal fault, a historical misnomer because this was once thought to be how faults normally occur.

Actually, most faults are produced by compressional forces, creating what is called a reversed fault, with one side of the fault pushed above the other side along a vertical or inclined plane. The great 1964 Alaskan earthquake produced as much as 50 feet of vertical displacement, forming

Figure 89 The Lewis thrust fault at the south end of Glacier National Park, Montana. Photo by M. R. Mudge, courtesy of USGS

a high scarp along the fault zone. If the reverse fault plane is nearly flat and the movement is mainly horizontal for great distances, it results in a thrust fault (Figure 89). A thrust fault occurs when a highly compressed plate shears so that one section is lifted over another. The overthrust belt from Canada to Arizona is an example of this type of faulting.

If a large block of crust bounded by normal faults is downdropped, it produces a long trenchlike structure called a graben. If a large block of crust bounded by reverse faults is uplifted, it produces a long ridgelike structure called a horst (Figure 90). Grabens and horsts are often found in association, forming long parallel mountain ranges and deep valleys like the Great East African Rift, Germany's Rhine Valley, the Dead Sea Valley in Israel and the Rio Grande Rift in the American Southwest. If a fault is a combination of vertical and horizontal movements, it forms a complex system known as an oblique fault.

Besides the San Andreas Fault, the United States is crisscrossed by numerous other faults, mostly associated with mountain ranges. Most states lie in regions that are classified as having moderate to major seismic risk. The Basin and Range province of southern Oregon, Nevada, western Utah, southeastern California, and southern Arizona and New

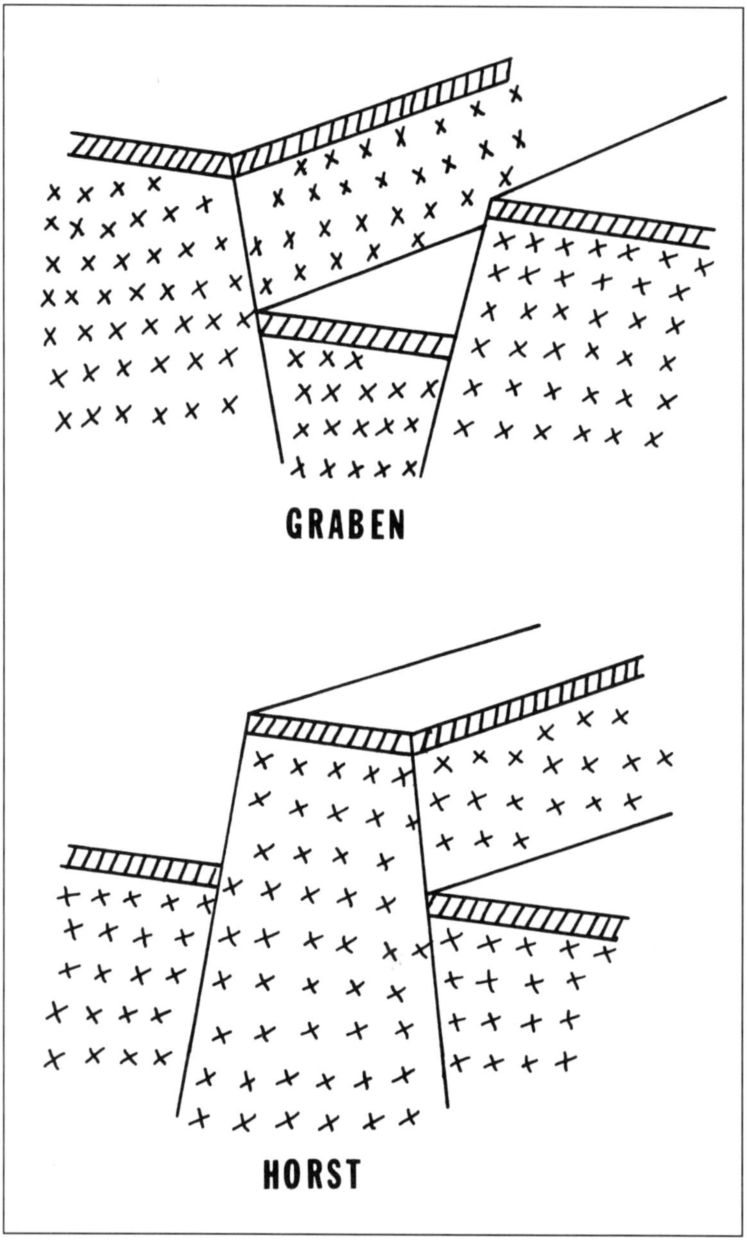

GRABEN

HORST

Figure 90 Graben and horst faults result from downdropped and upthrusted blocks.

Figure 91 Mount Moran in the Grand Tetons, Teton County, Wyoming. Photo by George A. Grant, courtesy of National Park Service

Mexico consists of numerous fault-block mountain ranges that are bounded by high-angle normal faults. The crust in this region is broken into hundreds of pieces that have been tilted and raised nearly a mile above the basin, forming nearly parallel mountain ranges 50 miles or more long. The region is literally being stretched due to the weakening of the crust by a series of downdropped blocks. About 15 million years ago, the locations presently occupied by Reno, Nevada and Salt Lake City, Utah were 200 to 300 miles closer together than they are today, due to this stretching.

The northern Rockies were created by the same mechanism of upthrusting connected with plate collision as the Andes of Central and South America. The Tetons of western Wyoming, one of the most spectacular ranges (Figure 91), were upfaulted along the eastern flank and downfaulted to the west. The Wasatch Range of north-central Utah and southern Idaho is an example of a north-trending series of normal faults, one below the other that extend for 80 miles with a net slip along the west side of as much as 18,000 feet.

The upper Mississippi and Ohio River valleys suffer frequent earthquakes, and the northeast-trending New Madrid Fault and its associated faults are responsible for three major earthquakes and numerous tremors. The Appalachian Mountains were formed by folding, faulting and upwarping of sediments and have been the seat of many earthquakes. Along the eastern seaboard, major earthquakes have hit Boston, New York, Charleston and other areas since colonial days.

Figure 92 Collapsed freeway overpass by the February 9, 1971, San Fernando, California, earthquake.
Courtesy of USGS

EARTHQUAKES

Earthquakes are by far the most destructive, short-term natural forces on Earth (Figure 92). Damage arising from earthquakes is widespread, covering thousands of square miles. Not only are entire cities destroyed, but earthquakes completely change the structure of the landscape in the affected region. They can produce tall, steep-banked scarps and cause massive landslides that carry away huge blocks of earth.

Hundreds of thousands of earthquakes occur every year (Table 11), but fortunately only a few are destructive. Since the turn of this century, the world has averaged about 18 major earthquakes of magnitudes 7.0 or larger per year. However, during the 1980s only 11 such shocks occurred on average per year. As for great earthquakes with magnitudes above 8.0, the century's average is 10 per decade, but the decade of the 1980s had only 4. It now appears, however, that the number of large earthquakes is again on the rise.

Vertical and horizontal offsets on the surface indicate that the crust is constantly readjusting itself. These movements are frequently associated with large fractures in the Earth. The greatest earthquakes are produced by sudden slippage along major faults, sometimes with offsets of 20 feet taking place in seconds. Most faults are associated with plate boundaries, and most earthquakes are generated in zones where plates are shearing past or abutting each other.

Where the plates interact, rocks at their edges are strained and deformed. This interaction can take place near the surface, where major earthquakes occur, or several hundred miles below, where one plate is subducted under another. Some faulting takes place so deep it leaves no surface expression.

TABLE 11 SUMMARY OF EARTHQUAKE PARAMETERS

Magnitude	Surface Wave Height (feet)	Length of Fault Affected (miles)	Diameter Area Quake Is Felt (miles)	Number of Quakes per Year
9	Largest earthquakes ever recorded are between 8 and 9			
8	300	500	750	1.5
7	30	25	500	15
6	3	5	280	150
5	0.3	1.9	190	1,500
4	0.03	0.8	100	15,000
3	0.003	0.3	20	150,000

Earthquakes are also associated with volcanic eruptions, but they are relatively mild compared to those created by faulting. Surprisingly, Antarctica and Greenland are devoid of significant earthquakes, probably because their massive ice sheets stabilize the faults and thus inhibit fault slip.

The total amount of slippage accumulated from earthquakes over time allows scientists to estimate the velocity at which the tectonic plates bounding the fault are moving past each other (Figure 93). By comparing this velocity with that computed by independent geologic, magnetic and geodetic evidence, it is possible to determine how much of the plate's relative motion causes earthquakes and how much produces aseismic slip, which is ground movement without producing earthquakes.

In areas like Chile, noted for some of the world's largest earthquakes, all motion between plates appears to be caused by earthquake slippage alone. The massive 1960 Chilean earthquake, the largest this century, took place along a 600-mile-long rupture through the South Chile subduction zone. Usually, the longer the section of fault that breaks, the larger is the earthquake.

Figure 93 Multiple fault traces and stream offsets on the San Andreas Fault, Carrizo Plains, California. Photo by R. E. Wallace, courtesy of USGS

Earthquakes also occur in so-called stable zones, although not nearly as frequently as they do at plate margins. The stable zones are generally associated with continental shields, which are composed of ancient granitic rocks lying in the interiors of the continents and account for nearly two thirds of all continental crust. When earthquakes occur in these regions, they might result from the weakening of the crust by compressive forces that originate at plate edges.

The underlying crust might also have been weakened by previous tectonic activity, including old faults and ancient mountain belts. Failed rift systems, where spreading centers did not fully develop, are responsible for faults such as the New Madrid in the central United States, which triggered three extremely large earthquakes in the winter of 1811–1812. Furthermore, the strong rock of plate interiors transmits seismic waves much more efficiently than the broken-up crust near plate boundaries. Therefore, earthquakes in these regions are felt over a much wider area. As an example, the New Madrid earthquake rang church bells as far away as Boston.

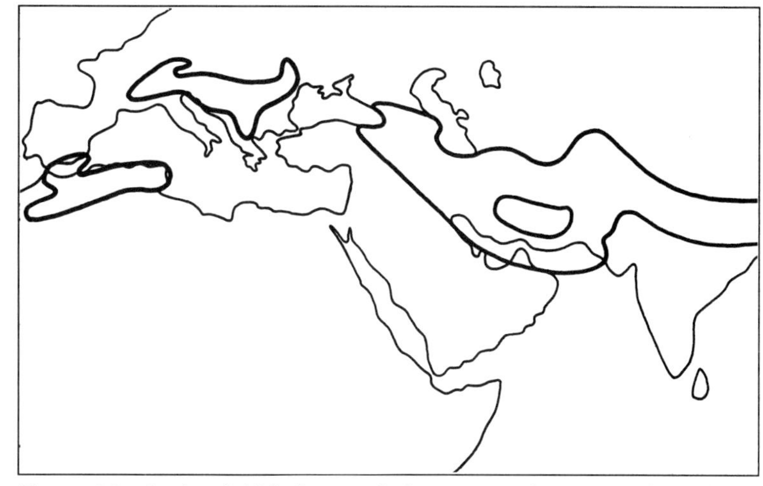

Figure 94 Active fold belts result from crustal compression where continental tectonic plates collide, such as the collision of Africa and Eurasia.

Earthquakes not associated with surface faults occur under folds, which do not rupture the Earth's surface. These earthquakes occur in many of the world's major fold belts that raise mountain chains, such as those bordering the Mediterranean Sea (Figure 94). During this last century, large-fold earthquakes have occurred in Japan, Argentina, New Zealand, Iran and Pakistan. Most of these earthquakes appear to have taken place under young anticlines, which are upturned strata less than several million years old, because folds are actually the geologic product of successive earthquakes and not by slow creep as was once thought.

The Pacific rim, the Mediterranean and central Asia experience most of the world's earthquakes. Nearly all earthquakes are concentrated in a few narrow zones that wind around the globe and are associated with plate boundaries. The greatest amount of seismic energy is released along a path located near the outer edge of the Pacific Ocean, known as the circum-Pacific belt. Another major belt runs through the folded mountainous regions flanking the Mediterranean Sea and through Iran past the Himalayan Mountains into China.

A continuous belt also extends for thousands of miles through the world's oceans and coincides with spreading ridge systems. Earthquakes are also associated with terrestrial rift zones, which are among the most active seismic regions, responsible for destroying civilizations through the ages.

8

IGNEOUS ACTIVITY

The very first rocks to form on Earth were igneous, derived directly from molten magma originating from deep within the planet. New magma is generated when crustal rocks melt as they sink into the mantle at subduction zones. Magma also upwells from the asthenosphere at spreading ridges and from deep inside the mantle at hot spots. The magma slowly rises toward the surface to become the source of all volcanic and granitic activities.

This igneous activity continues to build up the continents by the addition of new rock material. Therefore, the crust of the Earth is continuously being rejuvenated, and the total mass of buoyant rocks is always preserved. The addition of new basalt on the ocean floor is responsible for the growth of lithospheric plates, upon which the continents ride. The movement of plates is also responsible for the geologic forces that shape the planet.

MOLTEN MAGMA

The shifting of lithospheric plates on the surface of the planet generates new crust continuously. Deep-ocean trenches created by descending plates accumulate large amounts of sediment, primarily from the adjacent continents. The continental shelf and slope contain thick deposits of sediment washed off the continents. When these sediments are caught between subducting oceanic crust and continental crust, they are subjected to strong deformation, shearing, heating and metamorphism.

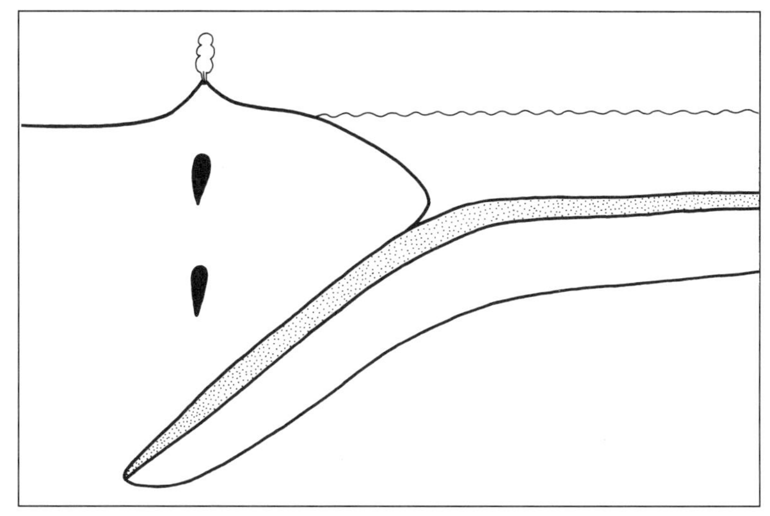

Figure 95 Crustal plates thrust deep inside the Earth melt and replenish volcanoes with new magma.

If the sediments are carried deep into the mantle, they melt in pockets called diapirs. The diapirs rise toward the surface to form magma bodies, which become the sources of new igneous activity (Figure 95). Subduction zones are noted for their active volcanism, which builds chains of continental volcanoes and island arcs. These volcanoes produce a fine-grained, gray rock known as andesite, named for the Andes Mountains of South America. Andesite differs in composition and texture from the upwelling basaltic magma of spreading ridges. It contains much more silica, which indicates a deep-seated source, possibly as deep as 70 miles below the surface.

Magma might also derive from partial melting of subducted oceanic crust, with the heat generated by the shearing action at the top of the descending plate. Convective motions in the wedge of athenosphere caught between the descending oceanic plate and the continental plate forces material upward, where it melts due to the lowered pressure.

The trenches, where plates dive into the mantle, are regions of low heat flow and high gravity because of the subduction of cool, dense lithosphere. Generally, the associated island arcs are regions of high heat flow and low gravity due to a high degree of volcanism. The back-arc basins behind island arcs are also regions of high heat flow due to the upwelling of magma from deep-seated sources.

About 80 percent of oceanic volcanism occurs at spreading ridges. Along the spreading ridges, magma wells up from the mantle and spews out onto the ocean floor. The spreading crustal plates grow by the steady accretion of solidifying magma to their edges. Over 1 square mile of new ocean crust, amounting to about 8 cubic miles of new basalt, is generated in this manner each year.

Seafloor spreading is often described as a wound that never heals, with magma slowly oozing out of the mantle. However, there are times when gigantic flows erupt on the ocean floor with enough new basalt to pave the entire U.S. interstate highway system 10 times over. The magma also flows from isolated volcanic structures called seamounts that are strung out in chains across the interior of plates (Figure 96).

The mantle material that extrudes onto the surface is black basalt, which is rich in silicates of iron and magnesium. Most of the more than 500 active volcanoes in the world are entirely or predominantly basaltic. The magma from which basalt is formed originated in a zone of partial melting in the Earth's upper mantle more than 60 miles below the surface. The semi-molten rock at this depth is less dense than the surrounding mantle material and rises slowly toward the surface. As the magma ascends, the pressure decreases and more mantle material melts. Volatiles such as dissolved water and gases also aid in making the magma flow easily.

The rising magma contributes to the formation of shallow reservoirs or feeder pipes that are the immediate source for volcanic activity. The magma chambers closest to the surface are under spreading ridges, where the crust is only 6 miles or less thick. Large magma chambers exist under fast spreading ridges where the lithosphere is being created at a high rate, such as those in the Pacific, and narrow magma chambers exist under slow spreading ridges, such as those in the Atlantic. As the magma chamber swells with magma and begins to expand, the crest of the spreading ridge is pushed upward by the buoyant forces generated by the molten rock. The magma rises in narrow plumes that mushroom out along the spreading ridge, welling up as a passive response to plate divergence, somewhat like having the lid taken off a pressure cooker. Only the center of the plume is hot enough to rise all the way to the surface, however. If the whole plume were to erupt, it could build a massive volcano several miles high. Not all magma is extruded onto the ocean floor. Some solidifies within the conduits above the magma chamber and forms massive vertical sheets known as dikes, which resemble a deck of cards standing on end.

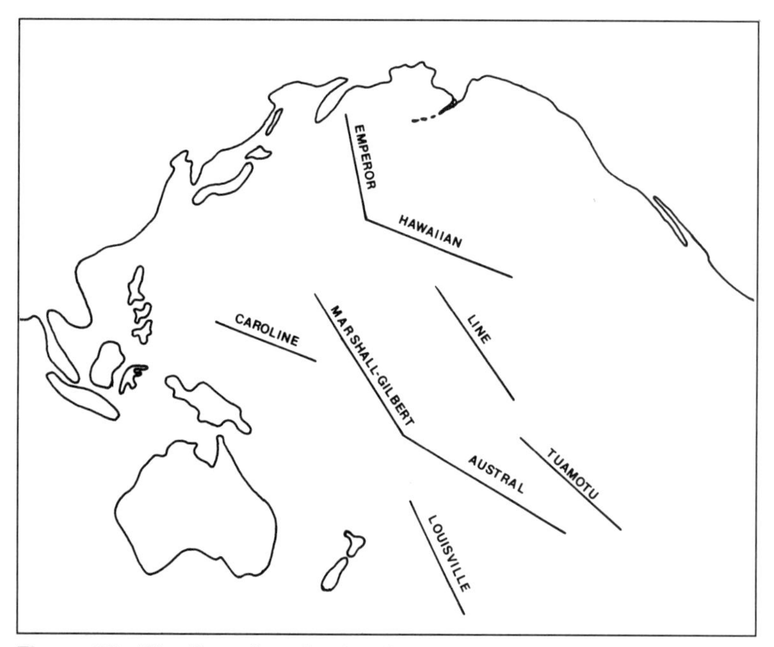

Figure 96 The linearity of volcanic seamounts on the Pacific plate point in the direction of plate travel.

Magmas of varied composition indicate the source materials as well as the depth within the mantle from which they came. Degrees of partial melting of mantle rocks, partial crystallization that enriches the melt with silica, and assimilation of a variety of crustal rocks in the mantle affect the ultimate composition of the magma. When the erupting magma rises toward the surface, it incorporates a variety of rock types along the way and changes composition, which is the major controlling factor in determining the type of eruption.

When the magma reaches the surface, it erupts a variety of gases, liquids and solids. Volcanic gases mostly consist of steam, carbon dioxide, sulfur dioxide and hydrochloric acid. The gases are dissolved in the magma and released as it rises toward the surface, and pressures decrease. The composition of the magma determines its viscosity and type of eruption, mild or explosive. If the magma is highly fluid and contains little dissolved gas when it reaches the surface, it flows from a volcanic vent or fissure as basaltic lava, and the eruption is usually quite mild, as with Hawaiian Island volcanoes (Figure 97).

If magma rising to the surface contains a large quantity of dissolved gases, it suddenly separates into liquid and bubbles. With decreasing pressure, the bubbles expand explosively and burst the surrounding fluid, which fractures the magma into fragments. The fragments are driven upward by the force of the expansion and hurled far above the volcano. The

Figure 97 The development of a 1,900-foot-high fountain during the December 17, 1959, eruption of Kilauea, Hawaii. Photo by D. H. Richter, courtesy of USGS

fragments cool and solidify during their flight and can range in size from large blocks weighing several tons to fine dust-size particles. The finer material is caught by the wind blowing across the eruption cloud and carried for long distances, sometimes completely around the world.

VOLCANIC ERUPTIONS

Most volcanoes are associated with crustal movements and occur on plate margins. When one crustal plate dives beneath another, the lighter rock component melts and rises toward the surface to provide magma for volcanoes and other igneous activity. A different type of volcano, called a hot-spot volcano, lies in the interiors of plates and arises from magma originating deep inside the mantle, possibly from just above the core. Volcanic islands such as the Hawaiian Islands are created as the Pacific plate travels over a hot spot, as though they were on a conveyer belt (Figure 98).

On the continents, hot spots leave a distinguishable trail of volcanoes. One such hot spot underlies Yellowstone National Park, Wyoming, and can be traced across the Snake River Plain in southern Idaho. Over the last 15 million years, the North American plate has traveled in a southwest direction over the hot spot, placing it under its temporary home at Yellowstone. During the last 2 million years, there were three major episodes of volcanic activity in the region. The volcanic eruptions that created the huge Yellowstone caldera, which is about 45 miles long and about 25 miles wide, are counted among the greatest catastrophes of nature, and another major eruption is well overdue.

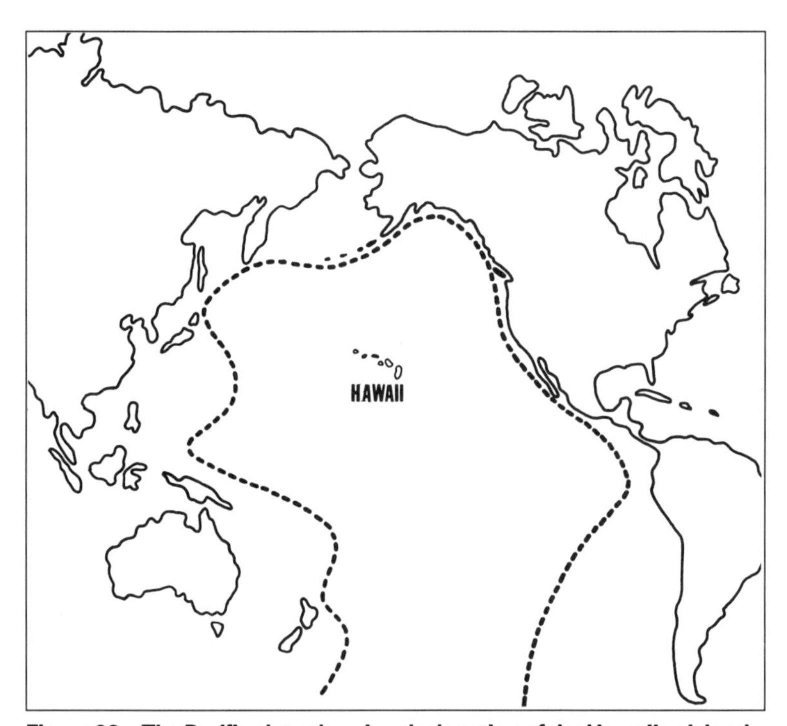

Figure 98 The Pacific plate showing the location of the Hawaiian Islands.

Subduction zone volcanoes such as those in the western Pacific and Indonesia are among the most explosive in the world, creating new islands and destroying old ones. The reason for their explosive nature is that the magma contains large amounts of volatiles in the form of water and gases. When the pressure is lifted as the magma reaches the surface, these volatiles are released explosively, fracturing the magma, which shoots out of the volcano like shotgun pellets.

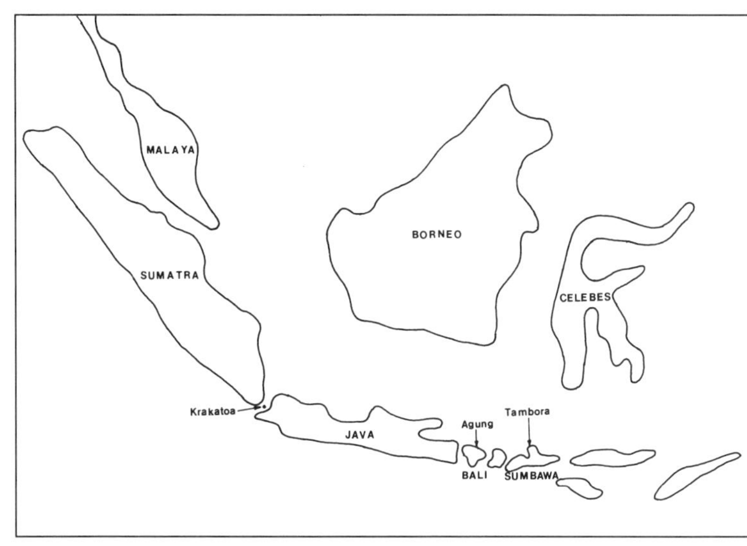

Figure 99 Location of major Indonesian volcanoes.

Indonesia is well known for its highly explosive volcanoes such as Tambora and Krakatoa (Figure 99), which caused the greatest eruptions in modern history (Table 12). The 1982 eruption of Galunggung and the 1983 eruption of Una Una created thick ash clouds that grounded aircraft. Alaskan volcanoes such as Mounts Katmai and Augustine, which are noted for their gigantic ash eruptions, resulted from the subduction of the Pacific plate at the Aleutian trench. The June 1991 eruption of Mount Pinatubo, Philippines, possibly the largest of the century, killed 700 people and left tens of thousands homeless.

The Cascade Range of the Pacific Northwest consists of a chain of powerful volcanoes from northern California to Canada. They are associ-

TABLE 12 THE 10 MOST DEADLY VOLCANOES

Date	Volcano	Area	Death Toll
A.D. 79	Vesuvius	Pompeii, Italy	16,000
1669	Etna	Sicily	20,000
1815	Tambora	Sumbawa, Indonesia	12,000
1822	Galunggung	Java, Indonesia	4,000
1883	Krakatoa	Java, Indonesia	36,000
1902	La Soufrière	St. Vincent Island	1,500
1902	Pelée	St. Pierre, Martinique	36,000
1902	Santa Maria	Guatemala	6,000
1919	Keluit	Java, Indonesia	5,500
1985	Nevado del Ruiz	Armero, Colombia	23,000

Figure 100 The May 18, 1980, eruption of Mount St. Helens, Washington. Photo by Austin Post, courtesy of USGS

ated with the Cascadia subduction zone, which is being overridden by the North American continent. The May 18, 1980, eruption of Mount St. Helens, whose blast devastated 200 square miles of national forest, is a good example of the explosive nature of these volcanoes (Figure 100). One of the dirtiest volcanoes of this century was El Chichon in southeastern Mexico, which began erupting on March 28, 1982. It sent volcanic dust clouds clear around the world, which had a major effect on the climate in the Northern Hemisphere. Mount Pinatubo had a similar effect 10 years later.

Volcanoes come in a variety of shapes and sizes. Cinder cones (Figure 101) are formed by explosive eruptions and are relatively short with steep slopes, usually less than 1,000 feet high. They are built up by accumulating layer upon layer of pumice, ash and other volcanic debris. Deep within the

Earth, viscous magma contains dissolved water, carbon dioxide and other gases. When the magma reaches the surface, the reduced pressure forces the gases out explosively, causing the volcano to spew its contents high into the air. The debris then falls back onto the volcano, building it upward as well as outward.

If a volcano erupts only basaltic lava from a central vent, it forms a shield volcano. Highly fluid molten rock from pools within the crater is violently squirted out in fiery fountains of lava or oozes out from a central vent. As the lava builds up in the center, it flows to the outer edge of the volcano in all directions, forming a dome-shaped structure when it cools and hardens. The slope on the volcano's flanks rises only a few degrees and no more than 10 degrees near the summit. The lava spreads out to cover large areas, as much as 1,000 square miles. Mauna Loa on Hawaii is the largest shield volcano in the world, creating a great sloping dome that rises 13,675 feet above sea level (Figure 102).

If a volcano erupts both cinder and lava, it builds a composite volcano, also called a stratovolcano. The hardened plug in the throat of the volcano is blasted into small fragments by the buildup of pressure from trapped gases below. Along with molten rock, these fragments are sent aloft and fall back on the volcano's flanks as cinder and ash. The cinder layers are reinforced by layers of lava from milder eruptions, forming cones with a steep summit and steeply slop-

Figure 101 Heavy cinder activity during the July 25, 1943, eruption of the Paricutin Volcano, Michoacan, Mexico. Photo by W. F. Foshag, courtesy of USGS

Figure 102 The broad shield volcano Mauna Loa built most of the island of Hawaii. Courtesy of USGS

ing flanks. These volcanoes are the tallest in the world, and often they end in a catastrophic collapse, forming a wide caldera. Most calderas form when a volcano loses its support and collapses into a partially emptied magma chamber. Calderas also form when a volcano blows off its upper peak, leaving behind a broad crater. If the floor of a caldera slowly domes up due to the resurgence of magma into a large chamber, it forms a resurgent caldera; Yellowstone is such a caldera.

At the summit of most volcanoes is a steep-walled depression, or crater. The crater is connected to the magma chamber by a conduit or vent. When fluid magma moves up the pipe, it is stored in the crater until it fills and overflows. During periods of inactivity, back flow can completely drain the crater. Highly viscous lava often forms a plug in the crater, which can slowly rise to form a huge spire or dome (Figure 103). Often, the lava is blown outward, greatly enlarging the crater.

Kimberlite pipes, named for the South African town of Kimberley, are cores of ancient extinct volcanic structures that extend deep into the upper mantle as much as 150 miles or more below the surface and have been exposed by erosion. Most known kimberlite pipes were emplaced during the Cretaceous period from 135 to 65 million years ago. They bring diamonds to the surface from the upper mantle and are mined extensively for these gems throughout Africa and other parts of the world. Most kimberlite pipes worth mining are cylindrical or slightly conical structures up to a mile across. Diamonds are created when pure carbon is subjected to extreme temperatures and pressures, conditions found deep inside the Earth.

VOLCANIC ROCK

Figure 103 Lava dome in the crater of Mount St. Helens. Photo by Jim Hughes, courtesy of USDA—Forest Service

The products of volcanic eruptions include gases, liquids and solids. The main factors controlling the physical nature of volcanic products are the viscosity of the magma, its water and gas content, the rate of emission and the environment of the vent. If, for example, the vent lies underwater or beneath a glacier, the same type of magma can produce entirely different rock types due to the different cooling rates.

Many subduction zone or island arc volcanoes have higher concentrations of gas in the upper parts of their magma chambers before they erupt. This is why Indonesian volcanoes such as Tambora and Krakatoa are so explosive. The eruption begins with the emission of pyroclastics, which literally means fire fragments (Figure 104). This is followed by thick, viscous lava flows. The texture of pyroclastics and lava is largely controlled by the number and size of gas bubble holes, called vesicles, formed in the erupting material. Pumice, the lightest of volcanic materials, contains the largest number of vesicles and can float on water.

Basalt, the densest volcanic rock, is formed at high temperatures and has practically no vesicles. It is the commonest rock formed from solidifying magma extruded on the surface of the Earth, moon and many other bodies in the Solar System. In some cases, especially on the ocean floor, basalt solidifies into elongated masses called pillow lava (Figure 105). As basalt lava cools on the surface, it shrinks, resulting in cracking or jointing. The cracks shoot vertically through the entire lava flow, breaking it into polygonal pillars or columns over a foot across.

Figure 104 Pyroclastic deposits at the base of Mount St. Helens Photo by P. Rowley, courtesy of USGS

All solid particles ejected into the air from volcanic eruptions are known as *tephra*, from the Greek word for "ash," a historical misnomer left over from the days when volcanoes were thought to arise from the burning of subterranean substances. Tephra includes an assortment of fragments from blocks the size of automobiles to dust-size material. It forms when molten rock containing dissolved gases rises through a conduit and suddenly separates into liquid and bubbles as it nears the surface. With decreasing pressure, the bubbles grow larger. If this event occurs near the orifice, a mass of

Figure 105 Pillow lava on Knight Island, Alaska. Photo by F. H. Moffit, courtesy of USGS

froth might spill out and flow down the sides of the volcano.

If this same reaction occurs deep down in the throat of the volcano, the bubbles expand explosively and burst the surrounding liquid, fracturing the magma into fragments. These are driven upward and hurled high above the volcano. The fragments cool and solidify during their flight. Often they whistle as they gyrate wildly through the air. Blobs of still-fluid magma, called volcanic bombs, might splatter the ground nearby (Figure 106). If they cool in flight, they form a variety of shapes, depending on how fast they are spinning. If the bombs are about the size of a nut, they form *lapilli*, from Latin meaning "little stones," which form strange gravel-like deposits after they land.

Tephra that flows down the slopes of a volcano under a layer of hot gases is called a *nuée ardente*, from the French for "glowing cloud." It flows streamlike near the ground and might follow existing river valleys for

Figure 106 A volcanic bomb, which fell on the east side of the Kilauea Volcano during the 1959–1960 eruption. Photo by R. T. Haugen, courtesy of National Park Service and USGS

miles, traveling upwards of 100 miles per hour. When the tephra cools and solidifies, it forms deposits called ash-flow tuffs that can cover an area of 1,000 square miles or more. Volcanoes also provide numerous samples of diverse welded tuffs, agglomerates and ignimbrites. Large ignimbrite sheets are composed of layers of welded or recrystallized volcanic ash.

Lava is molten magma that reaches the throat of a volcano or the top of a fissure vent without exploding into fragments and is able to flow onto the surface. The magma that produces lava is more fluid than that which produces tephra. This allows volatiles and gases to escape more easily and gives rise to much quieter and milder eruptions such as those of Kilauea on the main island of Hawaii. Lava, mostly composed of basalt, which contains only about 50 percent silica, is dark and quite fluid.

Ancient lava flows might contain clear, dark green or black natural glass called obsidian. Some lava flows might contain cavities filled with crystals

Figure 107 An aa lava flow entering the sea from the January 21, 1960, eruption of Kilauea, Hawaii. Photo by D. H. Richter, courtesy of USGS

called zeolites, meaning boiling stones, formed when water boiled away as the basalt cooled. Trachytes often contain large, well-shaped feldspar crystals that are aligned in the direction of the lava flow.

The outpourings of lava come in two general classes, which have Hawaiian names and are typical of Hawaiian eruptions. Pahoehoe or ropy lavas are highly fluid basalt flows formed when the surface of the flow congeals to form a thin plastic skin. Aa or blocky lavas form when viscous, subfluid molten rock presses forward, carrying a thick, brittle crust along with them. As the lava flows, it stresses the overriding crust, breaking it into rough, jagged blocks that are carried along with the flow in a disorganized mass (Figure 107).

Highly fluid lava moves fairly rapidly, especially down steep slopes. If a stream of lava hardens on the surface and the underlying magma continues to flow away, it forms a long cavern or tunnel called a lava cave that can reach tens of feet across and extend for hundreds of feet. The walls, roof and floor of the lava cave are often covered with stalactites and stalagmites composed of solidified lava.

Lava lakes, such as those on Mount Kilauea, are basalt flows that have been trapped in large pools that do not completely solidify. The magma that feeds the lakes rises from deep below the surface and is stored in a reservoir below the summit. The lava then erupts onto the surface and flows into a depression. The depth of the lakes can be substantial, as much as 400 feet. It takes a long time for lava lakes to cool and solidify, as much as one year for shallow lakes, to over 30 years for the deepest ones. Eventually, the natural dikes that channel the lava into the lake collapse, cutting off the lake from its source of lava, and it begins to solidify. Some lava lakes disappear completely as if the drain plug were pulled at the bottom of the crater.

GRANITIC INTRUSIVES

Magma bodies that invade the crust assimilate the surrounding rocks as they melt their way toward the surface. This produces two major classes of igneous rocks: intrusives, which are derived from the invasion of the crust by a magma body, and extrusives, which are derived from the eruption of magma onto the surface. Because their source materials are much the same, both types of rocks share similar chemical compositions but have different textures due to the difference in their cooling rates.

Magma that pours out onto the surface cools much faster than magma that remains in the crust and therefore gives rise to rocks with finer crystals. Intrusive bodies take a great deal of time to cool, perhaps 1 million years or more because the rocks they invade are good insulators and tend to hold the heat. The magma body is thus able to segregate into various compo-

Figure 108 Sheet joints formed in granitic rocks of the Sierra Nevada batholith. Photo by N. K. Huber, courtesy of USGS

nents, allowing large crystals to grow. Generally, the larger the magma body the longer it takes to cool and, consequently, the larger the crystals.

Intrusive magma bodies called plutons come in a variety of shapes and sizes. The largest plutons are batholiths, which are larger than 40 square miles and are usually longer than they are wide. Batholiths give rise to some of the world's major mountain ranges such as the Sierra Nevada in California (Figure 108), which is nearly 400 miles long but only about 50 miles wide, and the Idaho batholith, which is 250 miles long and 100 miles wide.

Batholiths consist of granitic rocks with large crystals, composed mostly of quartz, feldspar and mica. An intrusive magma body shaped much like a batholith but smaller than 40 square miles is called a stock. A stock might also be a projection of a larger batholith deeper down. Like a batholith, it is composed of coarse-grained granitic rocks.

If an intrusive magma body is tabular in shape and considerably longer than it is wide, it results in a dike. A dike forms when magma fluids occupy a large crack or fissure in the crust. Because dike rocks are usually harder than the surrounding material, they generally form long ridges when exposed by erosion (Figure 109). Sills are similar to dikes in their tabular form but are produced parallel to planes of weakness such as sedimentary beds. A special type of sill called a laccolith tends to bulge the overlying sediments upward, sometimes creating isolated mountain peaks in the middle of nowhere.

Figure 109 Shiprock and dike view from the south, San Juan County, New Mexico. Photo by H. E. Malde, courtesy of USGS

IGNEOUS ORE BODIES

For over 200 years there has been a lively debate among geologists on the origin of mineral ore deposits. The Neptunists hold that all mineral deposits were derived from water penetrating downward. The Plutonists, on the other hand, hold that mineralization was derived from the expulsion of magmatic volatiles migrating upward. These magmatic and hydrothermal processes are important in producing much of the world's wealth.

Mineral ore deposits form very slowly, taking several million years to create an ore rich enough to be suitable for mining. Copper, tin, lead and zinc ores are concentrated directly by magmatic activity, especially by the intrusion of magma bodies into the Earth's crust. These concentrations are formed as hydrothermal vein deposits, which are mineral fillings precipitated from hot waters percolating along underground fractures.

Around the turn of this century, geologists found that hot springs at Sulfur Bank, California, and at Steamboat Springs, Nevada (Figure 110), were depositing some of the same metal-sulfide compounds that are found in ore veins. Therefore, if the hot springs were depositing ore minerals at the surface, then hot water might be filling fractures in the rock with ore as it moves toward the surface.

After excavating the ground a few hundred yards from Steamboat Springs, the American mining geologist Waldemar Lindgren discovered that there were rocks with the texture and mineralogy of typical ore veins. He proved that many ore veins were formed by circulating hot water known as hydrothermal fluids. This concept vastly improved mineral exploration, because any evidence of hydrothermal alteration of rocks on the surface was enough to focus attention on the area. Unfortunately, only a small percentage of hydrothermal areas actually contains minable ore deposits. Therefore, some other processes must be at work.

Because most of the metals are found as sulfides, a source of sulfur and the chemistry that makes metal sulfides stable is needed. It is still unknown how hot water can carry enough metal to its place of deposition due to extremely low concentrations of these metals in solution. Either an ore deposit requires a huge amount of water over a very lengthy period, or some hot waters can carry more metal than what is observed on the surface.

It is possible that the rocks surrounding the magma chamber are the true source of the minerals found in hydrothermal veins. In this case, the volcanic rocks only act as a heat source that pumps groundwater into a giant circulating system. Cold, heavier water moves down and into the cooling volcanic rocks carrying trace quantities of valuable elements leached from the surrounding rocks. When heated by the cooling magma body, the water becomes less dense and rises into the fractured rocks above. After cooling and

Figure 110 Steam fumaroles at Steamboat Springs, Nevada. Courtesy of USGS

losing pressure, the water precipitates its mineral load into veins and moves down again to pick up another load of minerals.

A gigantic underground still is supplied with heat and some of the ingredients from magma chambers. As the magma cools, silicate minerals such as quartz crystallize first, leaving behind a concentration of other elements in a residual melt. Further cooling of the magma causes the rocks to shrink and crack, allowing the residual magmatic fluids to escape toward the surface and invade the surrounding rocks, where they form veins. Certain minerals precipitate over a wide range of temperatures and pressures, which is why they are commonly found together with one or two of the minerals predominating in high enough concentrations to make their mining profitable.

9

GLACIAL TERRAIN

Much of the landscape in the northern latitudes owes its unusual topography to massive ice sheets that swept down from the polar regions during the last ice age. The ice age was so pervasive that glaciers 2 miles or more thick enveloped much of upper North America and Eurasia. In some places, the crust was scraped completely clean of sediments, exposing the granitic basement rock below. In other areas, glacial sediments were deposited in massive heaps when the glaciers melted and retreated to the poles.

The ice age is still with us; we just live in a warmer period of it. Within another 2,000 years or so, the ice sheets could once again be on the move, wiping out everything in their paths. On their rampage, the ice sheets will bulldoze northern cities and shove their wreckage as far south as St. Louis.

THE DISCOVERY OF ICE

An old Norse tale recalls a period of time when winters were endless and the seas froze solid. Perhaps such lore was provoked by distant memories of a great ice age, when fully one third of the planet's land surface was covered by thick sheets of ice. It was only during the last two centuries that scientists have begun uncovering the geologic clues leading to the discovery of worldwide glaciation.

In many parts of the northern lands, huge blocks of granite weighing several thousand tons called erratic glacial boulders (Figure 111) were found strewn across the mountainous regions. Boulders found in the Jura Mountains could be traced to the Swiss Alps over 50 miles away. Most geologists of the late eighteenth century, however, believed that the boulders were swept out of the mountains by the Great Flood.

In 1760, the Swiss geologist Horace de Saussure discovered that downstream from the foot of a glacier the surfaces of projecting rocks along the glacial valley floor looked strikingly different from those high on the sides of the valley. The higher rocks were rough and jagged, whereas the lower ones were rounded, smooth and covered with numerous parallel scratches pointing down the valley. Rocks and boulders lay scattered about as though they were simply dumped there. From this observation, de Saussure concluded that glaciers had once extended far down the valley, grinding rocks on the valley floor as the ice advanced and receded.

Figure 111 Erratic glacial boulder in the Sierra Nevadas, Fresno County, California. Photo by G. K. Gilbert, courtesy of USGS

Some 30 years later, the Scottish geologist James Hutton described the Alps as having once been covered by a mass of ice consisting of immense glaciers that carried blocks of granite for great distances. However, most geologists of his day refused to believe that a river of solid ice with rocks embedded in it moved along the valley floor, grinding it down like a giant file as the glacier flowed back and forth. They were just as unconvinced that the glaciers were as widespread as Hutton predicted. Nor would they believe that glaciers were responsible for depositing isolated blocks of granite in the most unlikely places.

The Swiss civil engineer Ignatz Venetz strongly supported the ice age theory and in 1829 declared that alpine glaciers not only covered the Jura Mountains but extended far onto the European plain. He investigated marks on valley floors left behind by advancing glaciers as they ground their way down the mountainsides. He also visited several glaciers in various parts of the Swiss Alps. With these observations, Venetz was convinced that continental-size glaciers once existed over much of the northern lands.

The Swiss naturalist Louis Agassiz was perhaps the most ardent supporter of the glacial theory. He found substantial evidence that glacial ice masses had once blanketed the Swiss Mountains and covered the northern parts of Europe, Asia and America. In 1837, Agassiz led an expedition that included some of the most prestigious geologists in Europe to the Jura Mountains. On the valley floor, they found large areas of polished and deeply furrowed rocks miles from existing glaciers, and heaped rocks called moraines marked the extent of former glaciers. According to Agassiz, the valley was once buried by ice a mile or more thick. The glaciers

Figure 112 Terminal moraine at the margin of a glacier, Deschutes County, Oregon. Photo by I. C. Russell, courtesy of USGS

descended from the mountains, spread across most of northern Europe, and like giant bulldozers destroyed everything that got in their way.

Geologists in America gladly accepted Agassiz's glacial theories, for now they could explain such strange phenomena as gravel deposits known as lateral and terminal moraines (Figure 112), along with polished and striated rocks in the Northeast as well as other parts of the country. Long Island and Cape Cod were built entirely of thick moraines. Rocks in western New York State were polished and striated similar to those created by glaciers in the Jura Mountains. After immigrating to North America, Agassiz found that virtually all of the continent north of the Ohio and Missouri rivers, had once been glaciated. Since then, geologists have found evidence of several earlier ice ages, each rapidly following another (Figure 113).

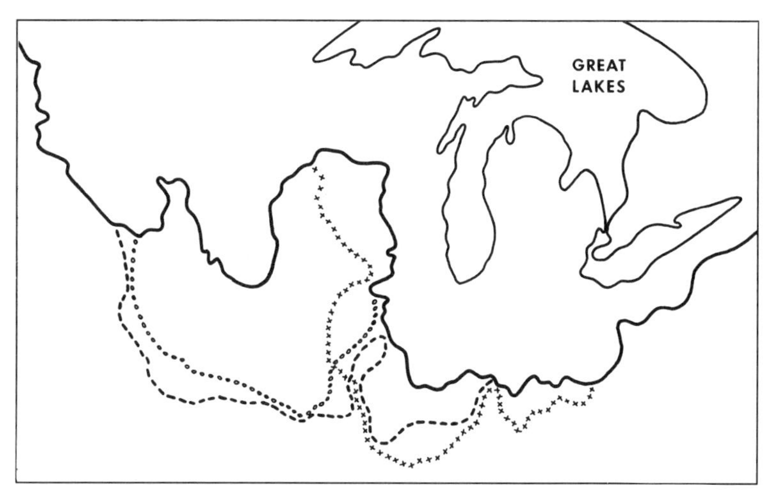

Figure 113 Maximum glacial advances in North America from the youngest to the oldest: Wisconsin (solid line), Illinoian (crosses), Kansan (dashes), Nebraskan (circles).

THE ICE AGE

The Pleistocene epoch, which began about 2.4 million years ago, witnessed a progression of ice ages, each followed by a short interglacial period similar to the one we are living in. The number of diatoms, which are one-celled algae with a shell made of silica (Figure 114), sharply declined in the Antarctic surface waters 2.4 million years ago when sea ice reached its maximum northern extent, shading the algae below. Without sunlight for photosynthesis, the diatoms simply vanished. Their disappearance also marks the initiation of the Pleistocene glaciation in the Northern Hemisphere.

The last ice age began about 100,000 years ago, intensified about 75,000 years ago, peaked about 18,000 years ago and retreated about 10,000 years ago. The ice sheets appear to have taken much longer to achieve their maximum extent than they took to recede to a comparably insignificant amount of ice in the polar regions today. In only a geologic moment in time, the ice sheets suddenly collapsed and rapidly disappeared as the global climate began to warm.

Because it was difficult to obtain reliable dates for the ice ages, it was impossible to fully test glacial theories because each succeeding ice age tended to erase the evidence of the one before. However, indirect evidence was found by studying coral terraces in the tropics. During the last ice age, about 5 percent of all the Earth's water was locked up in glacial ice. This resulted in an appreciable lowering of the sea and a significant expansion of the land area (Figure 115). Coral living in warm, shallow waters fluctuates in height in response to changing sea levels. Falling sea levels from the buildup of glacial ice culminated in the erosion of the coral reef. When the sea level rose after the glaciers melted, new coral grew on top of the old, forming a coral terrace. Alternating sea level changes corresponding to the waxing and waning of the glaciers produced a staircase-like structure of coral. The ages of the coral terraces were determined by radiometric dating techniques, which indicated the dates of the glacial events.

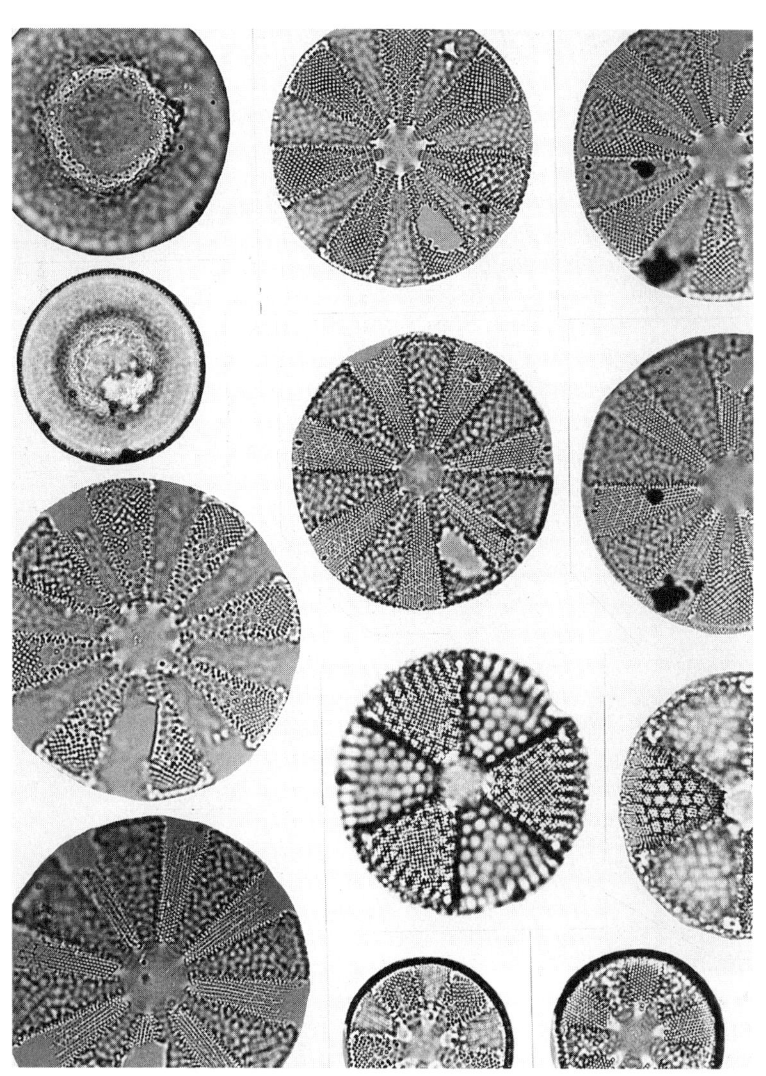

Figure 114 Miocene-age diatoms from the Choptank Formation, Calvert County, Maryland. Photo by G. W. Andrews, courtesy of USGS

During the height of the last glaciation, ice piled up 10,000 feet or more over Canada, Greenland and northern Europe. In North America, there were two main glacial centers. The largest ice sheet, the Laurentide, covered an area of 5 million square miles. It spread out from Hudson Bay and reached northward into the Arctic Ocean and southward to bury all of eastern Canada, New England and much of the rest of the northern half of the midwestern United States. A smaller ice sheet called the Cordilleran originated in the Canadian

Rockies and engulfed western Canada, parts of Alaska and small portions of the northwestern United States.

There were also two major glacial centers in Europe. The largest ice sheet called the Fennoscandian radiated from northern Scandinavia and covered most of Great Britain as far south as London and large parts of northern Germany, Poland and European Russia. A smaller ice sheet called the Alpine was centered in the Swiss Alps and covered parts of Austria, Italy, France and southern Germany. In Asia, ice sheets occupied the Himalayas and parts of Siberia.

In the Southern Hemisphere, small ice sheets grew in the mountains of Australia, New Zealand and the Andes of South America. Elsewhere, alpine glaciers existed on mountains that are presently ice-free. Only Antarctica had a major ice sheet, whose area increased by about 10 percent over its present size. The excess ice flowed into the sea, where it calved off to form icebergs (Figure 116). During the peak of the last ice age, icebergs covered about half the area of the oceans. Their light color reflected a great deal of sunlight into space, which maintained cool global temperatures and allowed the glaciers to continue growing. Large numbers of icebergs calving off the glaciers entering the sea also acted like giant ice cubes, which significantly lowered ocean surface temperatures.

Figure 115 Extended shoreline at the height of the last ice age.

At the height of the last ice age, the ocean fell by as much as 400 feet. An estimated 10 million cubic miles of water were incorporated into the continen-

Figure 116 Large iceberg along the coast of Grahm Land, Antarctic Peninsula. Photo by W. R. Curtsinger, courtesy of U.S. Navy

tal ice sheets, which covered about a third of the land surface with ice that was three times as large as its present volume. The lowered sea level extended the coastline of the eastern seaboard of the United States about halfway to the edge of the continental shelf, which lies more than 60 miles eastward of the present shoreline.

On the continental shelf off the eastern United States, a step on the ocean floor was traced for nearly 200 miles. It appears to represent the former ice age coastline, which is now submerged in several hundred feet of water. Submarine canyons carved into bedrock 200 feet below sea level can be traced to rivers on the continent. These submerged valleys were carved by rivers emptying into the sea when the sea level was lowered dramatically during the last ice age. This added more dry land to the continents and connected landmasses such as Asia and North America at the Bering Strait, which was instrumental to the migration of animals, including man.

The lowered temperatures also evaporate less seawater, reducing the average amount of precipitation. Since very little melting took place in the cooler summers, only a minimal amount of snowfall was needed to sustain the ice sheets. The lower precipitation levels also increased the size of deserts in many parts of the world. Desert winds were fierce and produced tremendous dust storms. So much dust was suspended in the atmosphere that it blocked out the sun, helping to make climatic conditions colder than normal. Most of the deposits of windblown sand in the central United States were laid during the Pleistocene ice ages.

Near the end of the last ice age, about a third of the ice melted between 16,000 and 13,000 years ago, when average global temperatures increased by about 5 degrees Celsius to nearly present-day levels. Around 13,000 years ago, a gigantic ice dam on the border between Idaho and Montana held back a huge lake hundreds of miles wide and up to 2,000 feet deep. When the dam suddenly burst and the waters gushed toward the Pacific Ocean, they carved one of the strangest landscapes on the planet, known as the Scablands (Figure 117). Lake Agassiz, which formed in a bedrock depression at the edge of the retreating ice sheet in south Manitoba, Canada, was a vast reservoir of meltwater much larger than any of the existing Great Lakes.

Figure 117 The Grand Coulee Dam on the Columbia River, in northeast Washington, showing tortured Scablands terrain. Courtesy of USGS

When the North American ice sheet began to retreat, its meltwater flowed down the Mississippi River. Sometimes, huge lakes of meltwater trapped beneath the ice sheet broke through and flowed in torrents down to the Gulf of Mexico and the Atlantic Ocean. While flowing under the ice, water surged in vast turbulent sheets that scoured deep

Figure 118 Niagara Falls on the border between New York and Ontario has been retreating at a rate of 2 to 4 feet per year. Photo by J. R. Balsley, courtesy of USGS

grooves in the surface, forming steep ridges carved out of the bedrock. Each flood lasted until the weight of the ice sheet once again shut off the outlet of the covered lake. Then another massive surge of meltwater broke loose to sculpt the landscape further.

After the ice sheet retreated beyond the Great Lakes, which were themselves carved out by the glaciers, the meltwater took a different route down the St. Lawrence River and the cold waters entered the North Atlantic. The rapid melting of the glaciers culminated in the extinction of microscopic organisms called foraminifera when a torrent of meltwater and icebergs spilled into the North Atlantic. This formed a cold, freshwater lid on top of the ocean, significantly changing the salinity of the seawater. Also during this time, the Niagara River Falls began cutting its gorge, which has moved over 5 miles northward since the ice sheet began to retreat (Figure 118).

GLACIAL DEPOSITS

Geologic evidence indicates that there were at least four major periods of glaciation in Earth history (Table 13). Most of the evidence for extensive glaciation is found in moraines and tillites, which are deposits of glacial

TABLE 13 CHRONOLOGY OF THE MAJOR ICE AGES

Years Ago	Event
2 billion	First major ice age
700 million	The great Precambrian ice age
230 million	The great Permian ice age
230–65 million	Interval of warm and relatively uniform climate
65 million	Climate deteriorates, poles become much colder
30 million	First major glacial episode in Antarctica
15 million	Second major glacial episode in Antarctica
4 million	Ice covers the Arctic Ocean
2 million	First glacial episode in Northern Hemisphere
1 million	First major interglacial
100,000	Most recent glacial episode
20,000–18,000	Last glacial maximum
15,000–10,000	Melting of ice sheets
10,000–present	Present interglacial

rocks. Tillites are a mixture of boulders and pebbles in a clay matrix consolidated into solid rock. They were deposited by glacial ice and are known to exist on every continent. In the Lake Superior region of North America, tillites are 600 feet thick in places and range east to west for 1,000 miles. In northern Utah, tillites mount up to an impressive thickness of 12,000 feet and provide evidence for a quick succession of ice ages.

Similar tillites were found among Precambrian rocks in Norway, Greenland, China, India, southwest Africa and Australia. In Australia, Permian marine sediments were found interbedded with glacial deposits, and tillites were separated by seams of coal, indicating that periods of glaciation were interspersed with warm interglacial spells. In South Africa, the Karroo Series, consisting of a sequence of late Paleozoic tillites and coal beds, reaches a total thickness of 20,000 feet.

Much of the upper midwest and northeastern parts of the United States were overrun by thick glaciers during the last ice age. Many areas were eroded down to the granite bedrock, erasing the entire geologic history of the region. The power of glacial erosion is well demonstrated by deep-sided valleys carved out of mountain slopes by thick sheets of ice (Figure 119). The glacially derived sediments covered much of the landscape, burying older rocks under thick layers of till.

The most recent glacial period is the best studied of all the ice ages, because evidence of each preceding glaciation was erased by the following

Figure 119 Saskatchewan Glacier, showing eroded glacial valley, Alberta, Canada. Photo by H. E. Malde, courtesy of USGS

one as ice sheets eradicated much of the landscape. In many areas, the ice stripped off entire layers of sediment, leaving behind bare bedrock. In other areas, older deposits were buried under thick deposits of glacial till, forming elongated hillocks aligned in the same direction and called drumlins (Figure 120). Drumlins are tall and narrow at the upstream end of the glacier and slope to a low, broad tail. The hills appear in concentrated fields in North America, Scandinavia, Britain and other areas once covered by ice. Drumlin fields might contain as many as 10,000 knolls, which look like rows upon rows of eggs lying on their sides.

Rugged periglacial regions existed at the margins of the ice sheets (Figure 121). Periglacial processes sculpted features along the tip of the ice and were directly controlled by the glacier. Cold winds blowing off the ice sheets affected the climate of the glacial margins and helped create periglacial conditions. The zone was dominated by such processes as frost heaving, frost splitting and sorting, which created immense boulder fields out of what was once solid bedrock.

Figure 120 Drumlins in Saskatchewan Province, Canada. Photo by W. G. Pierce, courtesy of USGS

Long, sinuous sand deposits called eskers (Figure 122) were formed out of glacial debris from outwash streams. They are winding, steep-walled ridges up to 500 miles long but seldom more than 1,000 to 2,000 feet wide and 150 feet high. Eskers were probably created by streams running through tunnels beneath the ice sheet. When the ice melted, the old stream deposits were left standing as a ridge. Well-known esker areas are found in Maine, Canada, Sweden and Ireland.

Glacial varves in ancient lake bed deposits are alternating layers of silt and sand laid down annually in a lake below the outlet of a glacier. Each summer when the glacial ice melted, meltwater turbid with sediments discharged into the lake, where the sediments settled out differentially, forming a banded deposit. The varying widths of the varves were thought to represent stages in the solar cycle when an increase in sunspot activity warmed the climate slightly, melting more ice than usual.

Dotting much of the northern lands are glacial lakes developed from major pits excavated by the glaciers. Smaller lakes were formed when a large block of ice buried by glacial outwash sediments melted and left a depression called a kettle. The depressions are nearly circular, ranging from a few hundred feet to 10 miles or more in diameter, and up to 100 feet or more deep. Not all kettles contain water, however, and some dry kettles hold a stand of trees that gradually fall off below ground level.

The largest glacial lakes are the Great Lakes between the United States and Canada. They presently receive huge amounts of sediments derived from the continent, and the constant buildup continues to make the lakes shallower. In the future, the lakes will dry out entirely, becoming flat, featureless plains, until once again the ice sheets come and scour out the basins.

SURGE GLACIERS

Antarctica discharges over 1 trillion tons of ice into the surrounding seas annually, and the ice calves off to form icebergs. Behind a wall of mountains in the Transantarctic Range, rivers of ice slowly flow outward and down to the sea on all sides. The ice escapes through mountain valleys to

Figure 121 Antonelli Glacier showing rugged periglacial area, including recessional and other moraines. Photo by R. B. Colton, courtesy of USGS

Figure 122 Esker near Mallory, Oswego County, New York. Photo by C. R. Tuttle, courtesy of USGS

the ice-submerged archipelago of West Antarctica, and to the great ice shelves of the Ross and Weddell seas (Figure 123).

West Antarctica is traversed by ice streams that are several miles broad, and rivers of solid ice flow down mountain valleys to the sea. The banks and the interior portions of the ice streams are marked by deep crevasses. On the bottom lie muddy pools of meltwater that lubricate the ice streams, allowing them to glide smoothly along the valley floors.

Much of the ice in Antarctica accumulated during the last ice age, when the ice sheets expanded to about a tenth larger than their present size. The ice is not uniform but pervaded by internal layers. Normally, when placed under great stress, ice will shatter. But because of its large size, a glacier acts like a flowing viscous solid, creeping over the landscape at approximately half a mile per year. Friction at the base of the glacier is lost due to huge subglacial lakes and streams. During a glacial surge, the streams stop flowing and the water spreads out beneath the glacier. This watery under-

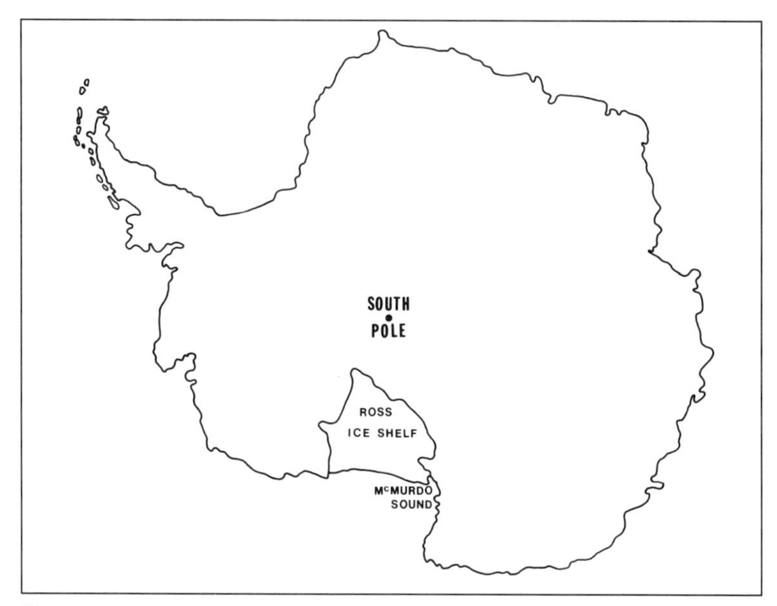

Figure 123 The huge ice shelf in West Antarctica.

coat acts like a lubricant, and large parts of the ice sheet surge along the ice streams toward the sea at speeds several times faster than normal.

Despite all this ice, Antarctica is virtually a desert with an average annual snowfall of only about 2 feet. This makes Antarctica one of the most impoverished, as well as one of the driest, regions on Earth. Dry valleys that run between McMurdo Sound and the Transantarctic Mountains were gouged out by local ice sheets (Fig-

Figure 124 The Taylor Glacier region, Victoria Land, Antarctica. Photo by W. B. Hamilton, courtesy of USGS

Figure 125 Tikke Glacier is one of nearly 200 surge glaciers in Alaska and adjacent Canada. Photo by Austin Post, courtesy of USGS

ure 124). They receive less than 4 inches of snowfall each year, most of which is blown away by strong winds. One interesting aspect about Antarctica is that experiments such as searching for traces of life have been conducted there for future excursions to Mars because the two landscapes are so similar.

There are some 200 surge glaciers in North America (Figure 125). During most of its lifetime, a surge glacier behaves normally, moving along at a couple of inches a day. However, at regular intervals of 10 to 100 years, the glaciers gallop forward up to 100 times faster than normal. A dramatic example is the Bruarjokull Glacier in Iceland, which in a single year advanced 5 miles, sometimes at an astonishing rate of 16 feet per hour.

A surge often progresses along a glacier like a great wave, proceeding from one section to another toward the sea. The reason for glacier surging has not yet been explained, although they might be influenced by the climate, volcanic heat and earthquakes. However, surge glaciers exist in the same regions as normal glaciers, often almost side by side. Another

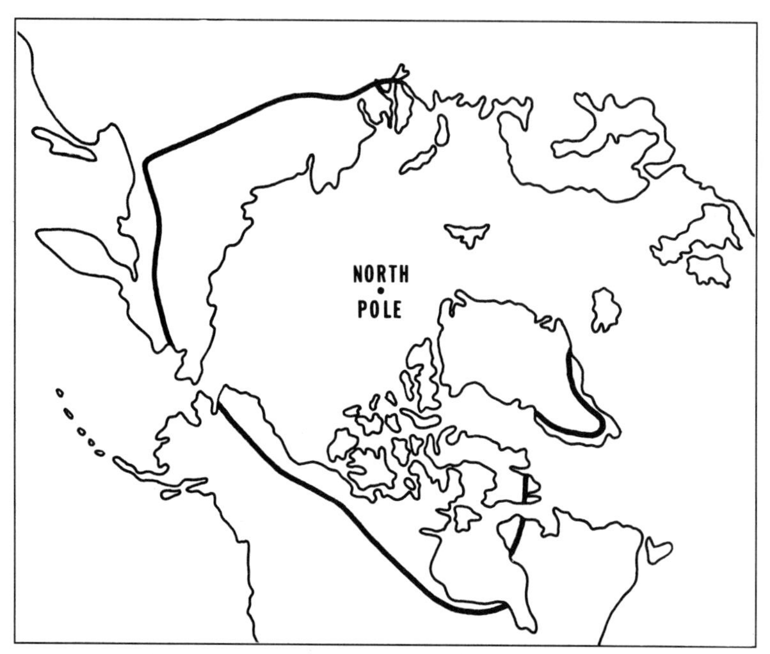

Figure 126 The arctic tundra line, north of which the ground remains frozen year round.

puzzling clue is the great 1964 Alaskan earthquake, which failed to produce more surges than normal.

PATTERNED GROUND

Among the most barren environments on Earth is the arctic tundra of North America and Eurasia. Tundra covers about 14 percent of the Earth's land surface in an irregular band that winds around the top of the world, north of the tree line and south of the permanent ice sheets (Figure 126). Most of the ground in the arctic tundra is permafrost and frozen year-round, and only the top few inches of the soil thaw during the short summers.

Although the ground is bathed in 24-hour sunlight during the summer, the soil temperature seldom reaches more than freezing. As it thaws, ice absorbs heat from soil particles around it. The arctic tundra is also one of the most fragile environments on Earth, and even small disturbances such as vehicle tracks from oil exploration activities can cause a great deal of damage lasting for decades.

In many parts of the tundra, soil and rocks are fashioned into strikingly beautiful and orderly patterns that have confounded geologists for centuries. Even on Mars, space probe images have revealed furrowed rings, polygonal fractures and ground-ice patterns of every description.

Every summer as the ground begins to thaw, the retreating snows in the arctic tundra unveil a bizarre assortment of rocks arranged in a honeycomb-like network, giving the landscape the appearance of a tiled floor (Figure 127). These patterns are found in most of the northern lands and alpine regions, where the soil is exposed to moisture and seasonal freezing and thawing. The polygons range in size from a few inches across when composed of small pebbles to 30 to 50 feet wide when large boulders form protective rings around mounds of soil.

The polygons were probably created by similar processes that cause frost heaving, which thrusts boulders upward through the soil, a major annoyance to northern farmers. Rocks have also been known to push through highway pavement, and fence posts have been shoved completely out of the ground by frost heaving. The boulders move through the soil by a pull from above or by a push from below. If the top of the rock freezes first, it is pulled upward by the expanding frozen soil. When the soil thaws, sediment gathers below the rock, which settles at a slightly higher level. The expanding frozen soil lying below can also heave the rock upward. After several frost-thaw cycles, the boulder finally comes to rest on the surface.

The regular polygonal patterned ground in the arctic regions might have been formed by the movement of soil of mixed composition upward toward

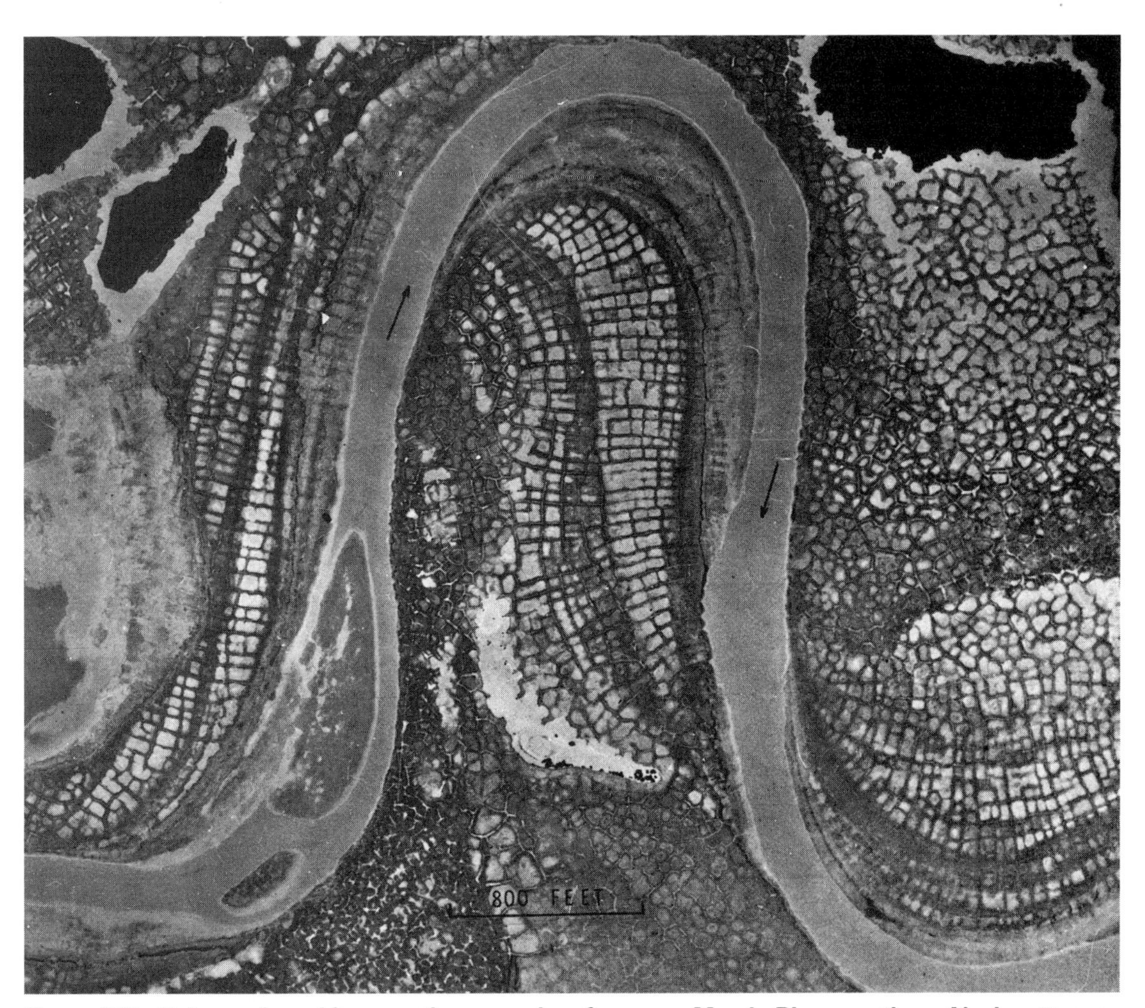

Figure 127 Polygonal markings on the ground surface near Meade River, northern Alaska. Photo by O. J. Ferrians, Jr., courtesy of USGS

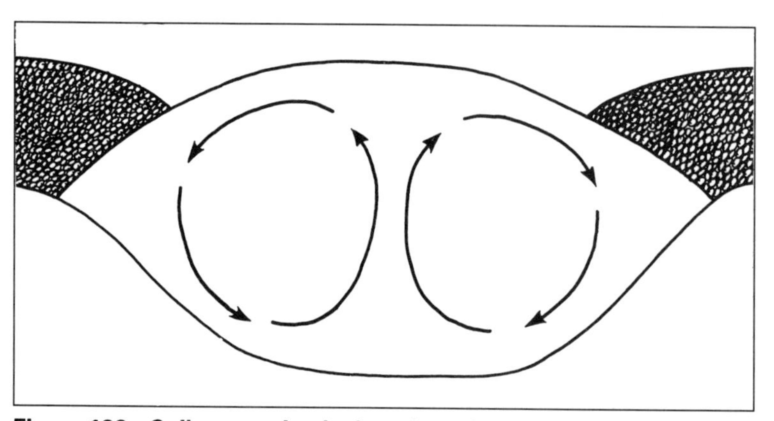

Figure 128 Soil convection is thought to be the process behind the creation of patterned ground.

the center of the mound and downward under the boulders, making the soil move in convective cells (Figure 128). The coarser material composed of gravel and boulders is gradually shoved radially outward from the central area, leaving the finer materials behind. The arrangement of rocks in this manner suggests that the soil is being churned by convection.

Other geometric designs in the arctic soil include steps, stripes and nets, which lie between the circles and polygons. These other forms can reach 150 feet in diameter. Relics of ancient surface patterns measuring up to 500 feet have been found in former permafrost regions. These among many other fascinating features presented in the Arctic makes the region one of the most interesting places on the face of the Earth.

10

UNIQUE STRUCTURES

The surface of the Earth is fashioned by various forces, including the interactions of crustal plates, gravitational impacts, erosion and collapse. These provide a large variety of unusual geologic structures, ranging from large-scale features such as rifts and faults, meteorite craters and collapse structures to individual sculptures in stone. Without these unique structures, the Earth's geology would be uninteresting to say the least, and the world would be a much duller place.

IMPACT STRUCTURES

When a large meteorite slams into the Earth, it kicks up an enormous amount of sediment and produces a deep crater (Figure 129). The finer material is lofted high into the atmosphere, while the coarse debris falls back around the perimeter of the crater, forming a high, steep-banked rim. A large meteorite traveling at high velocities completely disintegrates on impact, and the resulting explosion creates a crater some 20 times wider than the meteorite itself.

Upon striking the Earth, the meteorite sends a shock wave with pressures estimated at millions of atmospheres down into the rock and back up into the meteorite. As the meteorite burrows into the ground, it forces the rock aside while flattening itself in the process. It is then deflected like a

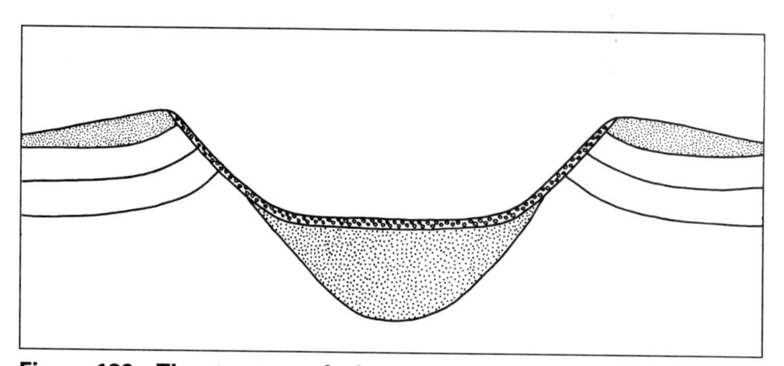

Figure 129 The structure of a large meteorite crater.

ricocheting bullet and thrown out of the crater, followed by a high-velocity spray of shock-melted meteoritic material along with melted and vaporized rock. As the spray continues to rise, it forms a rapidly expanding plume, similar to the mushroom cloud created by the detonation of an atomic bomb.

Not only are rocks shattered in the vicinity of the impact, but the shock wave passing through the ground induces shock metamorphism in the surrounding strata, resulting in changes in composition and crystal structure. The high temperatures developed by the force of the impact also fuses sediment into small glassy spherules called tektites. These are often spread evenly along the surface, forming wide strewn fields that can stretch halfway around the world. One example is mysterious pieces of glass strewn over the Western Desert of Egypt from a massive impact into the desert sands 30 million years ago. It scattered countless numbers of fist-sized, clear glass fragments all across the Libyan Desert.

TABLE 14 LOCATION OF MAJOR METEORITE CRATERS OR IMPACT STRUCTURES

Name	Location	Diameter (feet)
Al Umchaimin	Iraq	10,500
Amak	Aleutian Islands	200
Amguid	Sahara Desert	
Aouelloul	Western Sahara Desert	825
Baghdad	Iraq	650
Boxhole	Central Australia	500
Brent	Ontario, Canada	12,000
Campo del Cielo	Argentina	200
Chubb	Ungava, Canada	11,000
Crooked Creek	Missouri, USA	
Dalgaranga	Western Australia	250
Deep Bay	Saskatchewan, Canada	45,000

UNIQUE STRUCTURES

Name	Location	Diameter (feet)
Duckwater	Nevada, USA	250
Dzioua	Sahara Desert	
Flynn Creek	Tennessee, USA	10,000
Gulf of St. Lawrence	Canada	
Hagensfjord	Greenland	
Haviland	Kansas, USA	60
Henbury	Central Australia	650
Holleford	Ontario, Canada	8,000
Kentland Dome	Indiana, USA	3,000
Kofels	Austria	13,000
Lake Bosumtwi	Ghana	33,000
Manicouagan Reservoir	Quebec, Canada	200,000
Merewether	Labrador, Canada	500
Meteor Crater	Arizona, USA	4,000
Montagne Noire	France	
Mount Doreen	Central Australia	2,000
Murgab	Tadjikstan	250
New Quebec	Quebec, Canada	11,000
Nordlinger Ries	Germany	82,500
Odessa	Texas, USA	500
Pretoria Saltpan	South Africa	3,000
Serpent Mound	Ohio, USA	21,000
Sierra Madera	Texas, USA	6,500
Sikhote-Alin	Siberia	100
Steinheim	Germany	8,250
Talemzane	Algeria	6,000
Tenoumer	Western Sahara Desert	6,000
Vredefort	South Africa	130,000
Wells Creek	Tennessee, USA	16,000
Wolf Creek	Western Australia	3,000

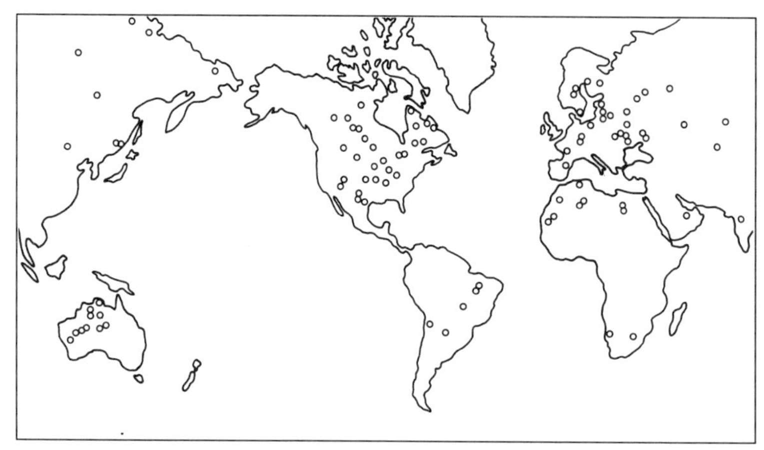

Figure 130 The location of major impact structures.

The world is dotted by more than 120 known impact structures (Figure 130; Table 14), which are large circular features created by the sudden shock from a large meteorite. They are generally circular or slightly oval in shape and are up to 50 miles or more wide. Some impact structures show distinct craters, and others display perhaps only subtle outlines of what were once

Figure 131 Meteor Crater near Winslow, Arizona. Courtesy of USGS

craters. The only evidence of their existence might be a circular disturbed zone, where rocks were altered by shock metamorphism. One of the best examples of this is a 10-mile-wide circular disturbed area in Tennessee called the Wells Creek structure.

In Quebec, Canada, the Manicouagan River and its tributaries form a reservoir around a roughly circular structure some 60 miles across. The structure consists of Precambrian rocks that were reworked by shock metamorphism from an impact of a large celestial body some 210 million years ago. The impact also coincides with the mass extinction that occurred at the end of the Triassic period, which suggests that extinctions in Earth history might be related to large meteorite impacts.

One of the most spectacular impact structures is Meteor Crater (Figure 131), located 15 miles west of Winslow, Arizona. It was first mistaken for a volcanic crater because volcanoes were once prevalent in the region. However, its distinctive appearance more closely resembles craters on the moon. Boreholes drilled into the center of the crater and on the south rim, which rises 135 feet above the desert floor, failed to locate the meteorite, indicating that it disintegrated upon impact. Indeed, several tons of metallic meteoritic debris were found scattered around the crater, evidence that the meteorite was the iron-nickle variety, measuring up to 200 feet wide and weighing about a million tons.

Most ancient meteorite impacts have been completely erased by the Earth's highly active erosional processes, including the action of rain, wind, glaciers, freeze-thaw cycles and organic activity. The Earth's moon and Mars retain their craters simply because they lack the weathering processes that have wiped out most signs of impacts on Earth.

The forces of erosion have leveled the tallest mountains and gouged out the deepest canyons. It is no wonder that the vast majority of craters do not escape these powerful weathering agents. The exceptions are craters in the deserts, which do not receive significant rainfall, or those in the arctic tundra regions, which remain mostly unchanged for ages. Ancient craters larger than 12 miles wide are usually deep enough so that the faint signs still remain. Circular depressions have even been found on the seafloor and are suspected of being impact structures created by large meteorites landing in the ocean. They must have created quite a splash falling into the sea, sending waves hundreds of feet high crashing down on nearby shores.

CRACKS IN THE CRUST

Slicing through the Earth are faults that often produce long, linear structures when exposed on the surface. The East African Rift Valley extends from the shores of Mozambique to the Red Sea, where it splits to form the

Figure 132 The San Andreas Fault in California. Courtesy of USGS

Afar Triangle in Ethiopia. The rift is a complex system of tensional faults, indicating that the continent is in the initial stages of being ripped apart.

Much of the area has been uplifted thousands of feet by an expanding mass of molten magma lying just beneath the crust. This heat source is responsible for the numerous hot springs and volcanoes along the rift valley. Some of the largest and oldest volcanoes in the world stand nearby, including Mounts Kenya and Kilimanjaro, which at 19,590 feet is the highest mountain in Africa.

Old extinct rift systems, where the spreading activity has ceased, or failed rifts, where a full-fledged spreading center did not develop, are sometimes overrun by continents. The western edge of North America has overrun the northern part of the now extinct Pacific rift system, creating the San Andreas Fault in California (Figure 132). A similar fault system, known as the Great Glen Fault, intersects Scotland from coast to coast and is causing the highlands to the north to slide past the lowlands to the south in a left-lateral direction. There is a string of deep lakes, including Loch Ness (of Loch Ness Monster fame), along the fault trace, which is marked by a belt of crusted, sheared rock up to a mile in width. Since the late Paleozoic, the fault has slipped as much as 60 miles, as shown by correlating geologic structures on both sides of the fault.

The frozen island of Iceland is split down the middle by the Mid-Atlantic Ridge, where the two plates that compose the Atlantic Basin and adjacent continents are being pulled apart (Figure 133). Upwelling magma at the spreading ridge adheres to the edges of the plates as well as to their bottom surfaces as the continents surrounding the Atlantic Basin separate at the expense of the Pacific Basin. The Pacific plates are being gobbled up by a ring of subduction zones, known as the Ring of Fire, which generates the most highly destructive volcanoes and earthquakes in the world.

Iceland itself is a broad volcanic plateau of the Mid-Atlantic Ridge that rises above sea level. A steep-sided, V-shaped valley runs up and down the island from north to south. It is flanked by numerous active volcanoes, which make Iceland one of the most volcanically active places on Earth. This accounts for a large amount of geothermal activity, providing heat and electricity for the residents. In a geologically brief moment, Iceland will move away from its source of magma, its volcanic activity will cease, and the island will become just another ice-covered rock.

The Grand Canyon of northern Arizona (Figure 134) is a 250-mile-long, 10-mile-wide, 1-mile-deep cut in the crust formed by the forces of uplift and erosion as the Colorado River sliced its way through half a billion years of accumulated sediments and Precambrian basement rock. Much of the canyon was not carved out by piecemeal erosion grain by grain but by catastrophic landsliding that tore away whole canyon walls.

PILLARS OF STONE

Out in the rugged West lie numerous monuments created when a resistant cap rock protected the sediments below while the surrounding landscape eroded, leaving behind tall pillars carved out of stone. Perhaps nowhere is this phenomenon better displayed than at Monument Valley on the border of Arizona and Utah near Four Corners. Instead of being crowded together like they are in other parts of the mile-high Colorado Plateau, the spires, mesas and ragged crags are widely scattered across the desert floor (Figure 135).

Some monuments stand alone as testaments to the unusual geologic activity that carved them out of the ancient rock. El Capitan Peak in Guadalupe Mountains National Park in west Texas is composed of a massive block of lime-

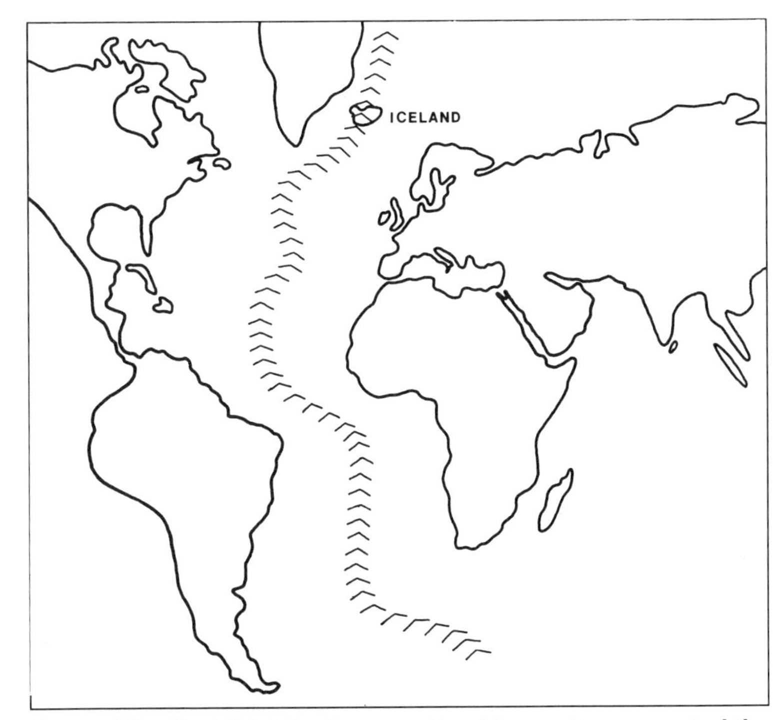

Figure 133 The Mid-Atlantic spreading ridge system separated the New World from the Old World. Note the position of Iceland.

stone rising high above its sloping flanks. Sometimes a long pinnacle called a chimney rock stands well above its surroundings. One such pinnacle is Chimney Rock National Historic Site near Scotts Bluff, Nebraska, which is named for the 800-foot bluff that stands alone in the middle of the prairie.

On the flanks of many monoclines, which are made up of steeply inclined sedimentary strata one below the other, rows of flatirons stand out from the rest of the formation. They look like old-fashioned steam irons standing on end and were formed by differential weathering of the strata. Flatirons are generously displayed in the steeply folded terrain of Utah and Wyoming. Figure 24 shows a series of flatirons in Alaska's Brooks Range.

Another type of folded strata is created when the crust is heaved upward, which is often due to salt tectonics. Since salt buried in the crust from ancient seabeds is lighter than the surrounding rocks, it slowly rises toward the surface, bulging the overlying strata upward. Often oil and gas are trapped in these structures, which explains why petroleum geologists spend a lot of time looking for salt domes. Upheaval Dome in Canyonlands National Park, Utah is perhaps the most striking example of a huge salt plug that heaved up the overlying strata into a huge bubblelike fold 3 miles wide and 1,500 feet high.

Figure 134 The Grand Canyon, Coconino County, Arizona. Courtesy of USGS

Bryce Canyon in southern Utah is fashioned out of colorful rocks similar to those that make up the Painted Desert farther to the south in Arizona. Uplift and erosion have carved a fantastic forest of pinnacles, spires and columns (Figure 136) created by a maze of ravines at the edge of a plateau. Similar structures can be found at Devil's Half Acre in central Wyoming.

In northeast Wyoming is Devil's Tower (Figure 137), which is an eroded volcanic plug that rises 1,300 feet above the surrounding prairie. It is composed of solidified magma that once filled the main conduit or feeder pipe of a volcano, and erosion has left the resistant rock standing alone. Along its flanks, the plug is broken by columnar jointing created by the shrinking magma as it cooled, creating fractures that shoot through its entire length. In northwest New Mexico lies a jagged monument called Shiprock, which rises 1,400 feet above the flat terrain. The volcanic neck is the remnant of volcanic eruptions that took place over 30 million years ago. Large dikes radiate outward in three directions like the spokes of a wheel.

Figure 135 Monument Valley from the top of Hunts Mesa, Navajo County, Arizona. A natural arch is in the left center of the photograph.
Photo by I. J. Witkind, courtesy of USGS

The devil seems to have been at work in California as well. Devil's Postpile southeast of Yosemite National Park contains rolls upon rolls of six-sided columns in a massive lava flow (Figure 138). As the lava cooled, it shrank, causing vertical cracking and jointing through the entire lava flow. Similar columns stretch across the bare cliffs of the Columbia River basalts, which buried much of Washington, Oregon and Idaho in basalt floods several thousand feet deep. Early legends have attributed the formation of columnar joints to supernatural forces, as reflected in the names of these sites.

The Canyonlands of Utah as well as other parts of the West are sprinkled with isolated pinnacles of sandstone called prairie dogs, named so because

Figure 136 Bryce Canyon from Inspiration Point, Utah. Photo by George A. Grant, courtesy of National Park Service

of the animal's penchant for standing up on its haunches to see above the grass on the lookout for predators. Several pinnacles might be clumped together into prairie dog towns. Sometimes erosion has played its usual tricks, and faces and other features can be seen with enough imagination.

Sometimes the statues come complete with their own headgear, including wide sombreros called Mexican hats. Perhaps the most prominent of these is found in southeast Utah north of Monument Valley, near a small town named, appropriately, Mexican Hat. The hat is formed out of a resistant cap rock that sits precariously on top of an eroded remnant.

Megalithic monuments such as Stonehenge in southern England and the great statues of Easter Island in the Pacific are among the most dramatic remains of prehistoric culture around the world. Since their discovery, there have been many theories put forward to explain their purpose and how they got there, evoking everything from extraterrestrial spaceships to supernatural forces. It appears that many of the hundreds of various circles of tall monoliths found in Europe were for astronomical purposes such as the telling of the seasons. The oldest monuments date to about 4000 B.C. and are often composed of exotic rock, having been hauled in from great distances. Such laborsome activity might have had something to do with the worship of stones.

In the petrified forests of Arizona and Wyoming are the remains of 200-foot-tall pines that were literally turned to stone. In Yellowstone National Park, the trees stand where they grew, preserved as stony stumps (Figure 139). A succession of forests has grown on top of each other, forming a layer cake of fossilized stumps that is 1,200 feet thick. In the Petrified Forest of Arizona are scattered trunks that were carried downstream by ancient floods and buried in the sediment. Groundwater percolating through the sediments replaced wood with silica, and erosion has exposed the petrified logs on the surface.

Submarine pillars exist, as well. In rapidly spreading rift systems such as the East Pacific Rise south of Baja California are hydrothermal vents on the ocean floor that build forests of tall chimneys often with branching pipes. They spew out large quantities of hot water blackened by sulfur compounds, and hence are called black smokers. Seawater seeping through the ocean crust is heated near magma chambers below the rifts and is expelled with considerable force through vents like undersea geysers (Figure 140). Living around these vents is a collection of some of the most bizarre creatures ever found on Earth. It is speculated that life originated around such vents and obtained all the necessary nutrients to survive from the Earth's hot interior.

Figure 137 Devil's Tower, Crook County, Wyoming, showing columns and talus. Photo by N. H. Darton, courtesy of USGS

Other marine forms called stromatolite mounds

Figure 138 Columnar joints in a massive basalt flow at Devil's Postpile National Monument, Madera County, California. Photo by H. R. Cornwall, courtesy of NASA

are composed of sediment that accumulated in concentric layers, forming tall, cabbagelike structures produced by algae living in intertidal zones. Their height is indicative of the height of the tides, which is controlled by the gravitational pull of the sun and moon. Ancient stromatolites grew taller than 30 feet, suggesting that at one time the moon was much closer to the Earth and exerted a stronger pull on the ocean, raising tides to substantial heights.

Back on land, mysterious Mima mounds, which are rounded piles of soil standing up to 10 feet high, are clustered in diverse parts of the world. They have puzzled scientists for more than 150 years and have possibly generated a greater variety of hypotheses than any other geologic feature. Mima mounds form in many earthquake-prone areas with markedly different climates, which suggests that they were caused by the vibrating ground during earthquakes. If a thin layer of soil rests on a section of hard rock, certain sections will vibrate more heavily than other sections where the mounds tend to pile up.

On an isolated ranch in central Washington State a large chunk of earth in the shape of a keyhole was discovered in 1984. It measured 7 by 10 feet wide and 2 feet high and weighed more than 1½ tons. But what was truly remarkable about this hunk of sod was that it lay about 75 feet from a hole in the ground with the exact same dimensions. The grass roots had been ripped out instead of cut as though the plug simply popped out of the ground. There were no signs of an explosion or other form of violent activity. However, the area was hit by a minor earthquake a few days before the plug was discovered.

Figure 139 Petrified tree trunks at Fossil Forest, Specimen Ridge, Yellowstone National Park, Wyoming. Photo by J. P. Iddings, courtesy of USGS

Earthquakes have been known to toss boulders and even people high into the air. One of the most dramatic examples of this occurred during the 1897 Assam earthquake in northeast India, which threw huge clods of earth in every direction, some of which landed with their roots pointing skyward. But none of these pieces of sod could compare with the colossal size of the Washington plug. Perhaps the energy of the earthquake was focused in such a way that it made the earth jump bodily out of its place and for an instant defy the laws of gravity.

HOLES IN THE ROCK

Differential weathering, which erodes rock at different rates due to differences in resistance, produces delicate arches like those found in Arches National Monument outside of Moab, Utah. The arches resulted partly from wind erosion of thick sandstone beds. Rainwater first loosens the sand near the surface, and wind removes the loose sand grains, causing abrasion that eats through the rock in ways similar to sandblasting. Wave erosion also cuts arches into sea cliffs.

Some arches are formed by the lateral erosion of a stream flowing around and eventually through the rock, forming a natural bridge (Figure 141). Rainbow Bridge in Utah north of Mexican Hat is the world's largest natural bridge. Many natural bridges formed in sandstone represent rock-shelter caves, where part of the roof has collapsed. Another type of natural bridge is formed when a large, detached block of rock falls or tilts so that it bridges the gap between two other blocks of rock. In limestone terrain, natural bridges are formed by tunnels excavated by groundwater solutions, resulting in a collapse of the tunnel roof. Natural Bridge in Virginia is the most famous example in the United States.

Caves are perhaps the most spectacular examples of the handiwork of groundwater. Over a lengthy period, the water dissolves great quantities of limestone, forming a system of tunnels, large rooms and galleries. The differences in shapes are caused by the geology, hydrology and structure of the rocks where caves form. Anvil Cave in Georgia has nearly 13 miles of passages spread over only 18 acres. Caves might also have more than one level, and some are completely vertical. The Mammoth Cave system in Kentucky is the longest in the world, with more than 300 miles of mapped passages and six known levels.

Smaller caves are fashioned out of dolomite and sandstone. Caves also develop in sea cliffs by the ceaseless pounding of the surf or by groundwater flowing through a limestone formation, which empties into the sea. Meltwater flowing out of a gla-

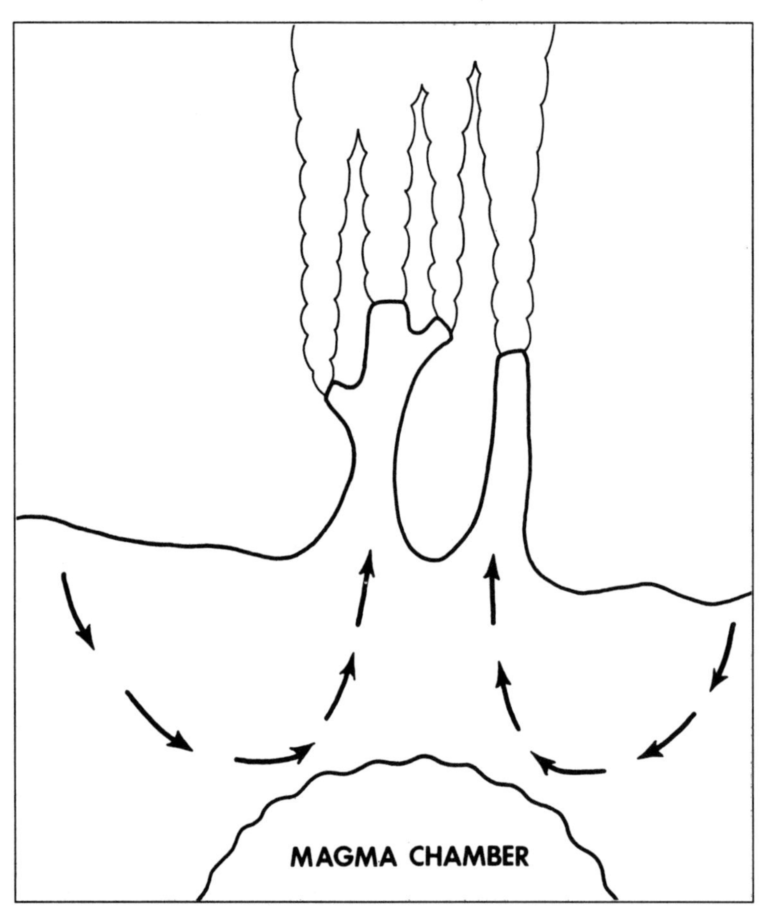

Figure 140 The operation of hydrothermal vents on the ocean floor.

Figure 141 Natural bridge in the Wasatch formation at the south end of Bryce Canyon National Park, Utah. Photo by R. G. Luedke, courtesy of USGS

cier forms ice caves that can be followed upstream for long distances. A lava flow that hardens on the surface while the molten rock in the interior continues to flow out forms a long lava cave. Many such caves are found in the volcanic terrain on Hawaii.

In limestone caves like Carlsbad Caverns in southeast New Mexico (Figure 142), long "icicles" of calcite grow by the precipitation of groundwater seeping through the rock. Those that form from the ceiling downward are called stalactites, and those that form from the floor upward are called stalagmites. A good mnemonic for telling the difference is to remember that stalactites hold tight to the ceiling, and stalagmites cling with all their might to the floor.

COLLAPSED STRUCTURES

If a volcanic crater is unusually large, one or more miles wide, it is called a caldera, which is Spanish for cauldron. Most calderas form when the summit of a volcano collapses after losing its support. Calderas are also formed when a volcano decapitates itself and blows off its upper peak, leaving behind a broad crater similar to one formed by the eruption of Mount St. Helens in 1980.

If dormant calderas fill with freshwater they form crater lakes. Crater Lake in Oregon (Figures 143A and 143B) was created when the upper 5,000 feet of the 12,000-foot composite cone of Mount Mazama collapsed 6,000 years ago and filled with rainwater and melting snow. The lake is 6 miles across and 2,000 feet deep, the deepest lake in North America. At one end of the lake projects a small volcanic peak known as Wizard Island.

Another crater lake called Nios, lying in northwest Cameroon in central Africa, was responsible for the deaths of over 1,700 people in 1986 when an earth tremor cracked open the deep lake bottom, releasing volcanic gases under high pressure. This created a gigantic bubble that exploded through

Figure 142 Hattin's Dome and the Temple of the Sun, Carlsbad Cavern, New Mexico, showing stalactites, stalagmites, columns and monuments. Photo by R. V. Davis, courtesy of USGS

the surface of the lake. The heavy gases swept down the hillside and spread out in a low-hanging blanket for over 3 miles downwind from the lake, asphyxiating its victims with a brew of deadly gases.

Hole in the Ground in the Cascade Mountains of Oregon is the site of a gigantic volcanic gas explosion that left a huge crater, resembling those at the Nevada Test Site from underground nuclear detonations. It is a perfectly circular pit several thousand feet across and has a rim several hundred feet above the surrounding terrain. For unknown reasons, most of the crater is devoid of vegetation.

Figure 143A Crater Lake in Klamath County, Oregon formed when the Mazama Volcano collapsed 6,000 years ago. Courtesy of USGS

Death Valley, the lowest point on the North American continent at 280 feet below sea level, was at one time several thousand feet higher. The area collapsed when the continental crust thinned due to extensive block faulting in the Basin and Range province, resulting in a series of north-south-trending mountain ranges. The entire Great Basin area is a remnant of a broad belt of mountains and high plateaus that collapsed as the crust was pulled apart during the formation of the Rocky Mountains.

Man-made collapse structures include areas of subsidence due to overpumping of groundwater or subterranean oil, which causes an aquifer or reservoir to compact, substantially lowering the level of surface sediments over a wide area. Collapse of underground coal mines in the eastern states has dropped surface sediments several feet (Figure 144), often taking with them houses built on top of the mines. If a limestone

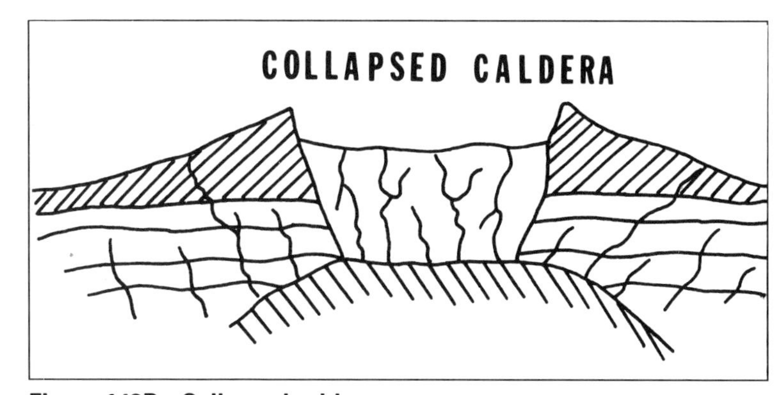

Figure 143B Collapsed caldera.

Figure 144 Subsidence depressions, pits and cracks above abandoned underground mine, Sheridan County, Wyoming. Photo by C. R. Dunrud, courtesy of USGS

Figure 145 A large sinkhole 520 feet long, 125 feet wide, and 60 feet deep collapsed under a house in Barton, Florida. Courtesy of USGS

formation is excavated by groundwater forming a large cavity just below the surface and the ground loses its support and caves in, it results in a sinkhole. Sinkholes are usually found in karst topography, which is characterized by sinkholes, caves, caverns and the like, and found especially in Florida and Alabama. Sometimes a sinkhole fills with water to form a small lake. If a sinkhole occurs in a residential area, entire houses might suddenly disappear (Figure 145).

GLOSSARY

aa lava	a lava that forms large, jagged, irregular blocks
agglomerate	a pyroclastic rock composed of consolidated volcanic fragments
alluvium	stream-deposited sediment
alpine glacier	a mountain glacier or a glacier in a mountain valley
andesite	an intermediate type of volcanic rock between basalt and rhyolite
anticline	folded sediments that slope downward away from a central axis
aquifer	a subterranean bed of sediments through which groundwater flows
arkose	a feldspar-rich sandstone
ash fall	the fallout of small, solid particles from a volcanic eruption cloud
asthenosphere	a layer of the upper mantle, roughly between 50 and 200 miles below the surface, that is more plastic than the rock above and below and might be in convective motion
basalt	a volcanic rock that is dark and usually quite fluid in the molten state

batholith	the largest of intrusive igneous bodies, at least 40 square miles on its uppermost surface
bedrock	solid layers of rock beneath younger materials
blueschist	metamorphosed rocks of subducted ocean crust exposed on land
breccia	a clastic sedimentary rock composed of angular fragments
caldera	a large pitlike depression at the summits of some volcanoes formed by great explosive activity followed by collapse
carbonaceous	describes substances containing carbon, namely sedimentary rocks such as limestone and certain types of meteorites
carbonate	describes minerals containing calcium carbonate such as limestone and dolostone
chert	an extremely hard, fine-grained quartz mineral
circum-Pacific belt	active seismic regions around the rim of the Pacific plate coinciding with the Ring of Fire
cirque	a glacial erosion feature, producing an ampitheater-like head of a glacial valley
conduit	a passageway leading from a reservoir of magma to the surface of the Earth through which volcanic products pass
conglomerate	a sedimentary rock composed of welded fine-grained and coarse-grained rock fragments
continent	a landmass composed of light, granitic rock that rides on denser rocks of the upper mantle
continental drift	the concept that the continents have been drifting across the surface of the Earth throughout geologic time
continental glacier	an ice sheet covering a portion of a continent
continental shelf	the offshore area of a continent in shallow sea

continental shield	ancient crustal rocks upon which the continents grew
continental slope	the transition from the continental margin to the deep-sea basin
convection	a circular, vertical flow of a fluid medium due to heating from below. As materials are heated, they become less dense and rise, whereas cooler, heavier materials sink
cordillera	a range of mountains that includes the Rockies, Cascades and Sierra Nevada in North America and the Andes in South America
core	the central part of a planet and consisting of a heavy iron-nickel alloy
correlation	the tracing of equivalent rock exposures over distance
craton	the stable interior of a continent, usually composed of the oldest rocks on the continent
crevasse	a deep fissure in the earth or a glacier
crust	the outer layers of a planet's or a moon's rocks
crustal plate	one of several plates comprising the Earth's surface rocks
diapir	the buoyant rise of a molten rock through heavier rock
dike	a tabular intrusive body that cuts across older strata
dolomite	a sedimentary rock formed by the replacement of calcium with magnesium in limestone
drumlin	a hill of glacial debris facing in the direction of glacial movement
East Pacific Rise	a midocean spreading center that runs north-south along the eastern side of the Pacific; the predominant location upon which the hot springs and black smokers have been discovered
elastic rebound theory	the theory that earthquakes depend on rock elasticity
eolian	describes a deposit of windblown sediment

epicenter	the point on the Earth's surface directly above the focus of an earthquake
erratic	a boulder glacially deposited far from its source
escarpment	a mountain wall caused by elevation of a block of land
esker	curved ridges of glacially deposited material
evaporite	the deposition of salt, anhydrite and gypsum from the evaporation in an enclosed basin of stranded seawater
extrusive	any igneous volcanic rock that is ejected onto the surface of a planet or moon
facies	an assemblage of rock units deposited in a certain environment
fault	a breaking of crustal rocks caused by earth movements
feldspar	the most abundant rocks in the Earth's crust composed of silicates of calcium, potassium and sodium
fissure	a large crack in the crust through which magma might escape from a volcano
fluvial	pertaining to being deposited by a river
focus	the point of origin of an earthquake; also called hypocenter
formation	a combination of rock units that can be traced over distance
fossil	any remains, impression or trace in rock of a plant or animal of a previous geologic age
frost heaving	the lifting of rocks to the surface by the expansion of freezing water
frost polygons	polygonal patterns of rocks by repeated freezing
gabbro	a dark, coarse-grained intrusive igneous rock
glacier	a thick mass of moving ice occurring where winter snowfall and ice formation exceed summer melting

gneiss	pronounced "nice," a banded, coarse-grained metamorphic rock with alternating layers of unlike minerals. It consists of essentially the same components as granite
Gondwana	a southern supercontinent of Paleozoic time, consisting of Africa, South America, India, Australia and Antarctica. It broke up into present continents during the Mesozoic era
granite	a coarse-grained, silica-rich rock consisting primarily of quartz and feldspars. It is the principal constituent of the continents and is believed to be derived from a molten state beneath the Earth's surface
graywacke	a poorly sorted sandstone with a clay matrix
greenstone	a green metamorphosed igneous rock of the Archean age
groundwater	the water derived from the atmosphere that percolates and circulates below the surface of the Earth
guyot	an undersea volcano that reached the surface of the ocean, whereupon its top was flattened by erosion. Later, subsidence caused the volcano to sink below the surface, preserving its flattop appearance
gypsum	a calcium sulfate mineral formed during the evaporation of brine pools
half-life	the time for one-half the atoms of a radioactive element to decay
halite	an evaporite deposit composed of common salt
hiatus	a break in geologic time due to a period of erosion or nondeposition of sedimentary rocks
horn	a peak on a mountain formed by glacial erosion
hot spot	a volcanic center that has no relation to a plate boundary location; an anomalous magma generation site in the mantle
hydrothermal	relating to the movement of hot water through the crust
hydrothermal deposit	a mineral ore deposit emplaced by hot groundwater

hypocenter	the point of origin of earthquakes; also called the focus
Iapetus Sea	a former sea that occupied a similar area as the present Atlantic Ocean prior to the assemblage of Pangaea
ice age	a period of time when large areas of the Earth were covered by glaciers
iceberg	a portion of a glacier broken off upon entering the sea
ice cap	a polar cover of ice and snow
igneous rocks	all rocks that have solidified from a molten state
ignimbrite	volcanic deposits created by ejections of incandescent solid particles
interglacial	a warming period between glacial periods
intrusive	any igneous body that has solidified in place below the surface of the Earth
island arc	volcanoes landward of a subduction zone, parallel to the trench, and above the melting zone of a subducting plate
isostasy	the idea that the crust is buoyant and rises or sinks depending on its weight
kettle	a depression in the ground caused by a buried block of glacial ice
kimberlite	a diamond-bearing volcanic material originating deep within the mantle
lahar	a mudflow of volcanic material on the flanks of a volcano
lapilli	small, solid pyroclastic fragments
Laurasia	the northern supercontinent of the Paleozoic consisting of North America, Europe and Asia
lava	molten magma after it has flowed out onto the surface
limestone	a sedimentary rock consisting mostly of calcite
liquefaction	the loss of support of sediments, which liquefy during an earthquake

lithosphere	a rigid outer layer of the mantle, typically about 60 miles thick. It is overridden by the continental and oceanic crusts and is divided into segments called plates
loess	pronounced LOW-ess, a thick deposit of fine-grained windblown sediment
magma	a molten rock material from within the Earth that, when hardened, forms igneous rocks. It is often generated during volcanic eruptions
magnetic field reversal	a reversal of the north-south polarity of the Earth's magnetic poles. This has occurred intermittently throughout geologic time
magnetite	a dark, iron-rich, strongly magnetic mineral, sometimes called a lodestone
magnetometer	a devise used to measure the intensity and direction of the magnetic field
magnitude scale	a scale for rating earthquake energy
mantle	the part of a planet below the crust and above the core, composed of dense iron-magnesium-rich rocks
maria	dark plains on the lunar surface caused by massive basalt floods
metamorphic rock	a rock crystallized from previous igneous, metamorphic or sedimentary rocks created under conditions of intense temperatures and pressures without melting
meteoritic crater	a depression in the crust produced by the bombardment of a meteorite
microearthquake	a small Earth tremor
Mid-Atlantic Ridge	the seafloor spreading ridge of volcanoes that marks the extensional edge of the North American and South American plates to the west and the Eurasian and African plates to the east
midocean ridge	a submarine ridge along a divergent plate boundary where a new ocean floor is created by the upwelling of mantle material

Moho	the boundary between the crust and mantle. Moho is short for Mohorovicic discontinuity, so called after the geologist who discovered it
moraine	a ridge of erosional debris deposited by the melting margin of a glacier
mountain roots	the deeper crustal layers under mountains
Neolithic	a period of new stone age 35,000 to 10,000 years ago
nonconformity	an unconformity in which sedimentary deposits overlie crystalline rocks
normal fault	a gravity fault in which one block of crust slides down another block of crust along a steeply tilted plane
nuée ardente	a volcanic pyroclastic eruption of hot ash and gas
ophiolite	rocks composed of masses of oceanic crust thrust onto continents by plate collisions
orogeny	a process of mountain building by tectonic activity
overthrust	a thrust fault in which one segment of crust overrides another segment of crust
pahoehoe lava	a lava that forms ropelike structures when cooled
paleomagnetism	the study of the Earth's magnetic field, including the position and polarity of the poles in the past
paleontology	the study of ancient life forms, based on the fossil record of plants and animals
Pangaea	an ancient supercontinent that included all the lands of the Earth
Panthalassa	the great world ocean that surrounded Pangaea
pegmatite	a very coarse grained igneous rock
peneplain	a ridge representing an original plain
peridotite	a coarse-grained igneous rock composed mainly of olivine and pyroxene and main constituent of the mantle
permeability	the ability to transfer fluid through cracks, pores and interconnected spaces within a rock

placer — a deposit of rocks left behind from a melting glacier; also, any ore deposit that is enriched by stream action

plate tectonics — the theory that accounts for the major features of the Earth's surface in terms of the interaction of lithospheric plates

pluton — an underground body of igneous rock younger than the rocks that surround it. It is formed where molten rock oozes into a space between older rocks

pumice — volcanic ejecta that is full of gas cavities and extremely light in weight

pyroclastic — describes the fragmental ejecta released explosively from a volcanic vent

quartz — a common igneous-rock-forming mineral of silicon dioxide

radiometric dating — the determination of how long an object has existed by chemical analysis of stable verses unstable radioactive elements

radionuclide — a radioactive element that is responsible for generating the Earth's internal heat

redbed — red-colored sedimentary rocks indicative of a terrestrial deposition

reef — the biological community that lives at the edge of an island or continent. Shells form a limestone deposit that is readily preserved in the geologic record

regression — a fall in sea level, exposing continental shelves to erosion

remnant magnetism — a permanently induced magnetic field in a rock

reversed magnetism — a geomagnetic field with a reverse polarity from that of the present one

rhyolite — a volcanic rock that is highly viscous in the molten state and usually ejected explosively as pyroclastics

rift valley — the center of an extensional spreading center, where continental or oceanic plate separation occurs

saltation	the movement of sand grains by wind or water
sandstone	a sedimentary rock consisting of sand grains cemented together
scarp	a steep slope formed by Earth movements
schist	a finely layered metamorphic rock that tends to split readily into thin flakes
seafloor spreading	the theory that the ocean floor is created by the separation of lithospheric plates along the midocean ridges, with new oceanic crust formed from mantle material that rises from the mantle to fill the rift
seamount	a submarine volcano
sedimentary rock	a rock composed of fragments cemented together
seismic sea wave	an ocean wave related to an undersea earthquake
seismometer	a detector of earthquake waves
shield	areas of the exposed Precambrian nucleus of a continent
shield volcano	a broad, low-lying volcanic cone built up by lava flows of low viscosity
siderophile	an element with an affinity toward iron
stratovolcano	an intermediate volcano characterized by a stratified structure from alternating emissions of lava and fragments
striation	scratches on bedrock made by rocks embedded in a moving glacier
stromatolite	a calcareous structure built by successive layers of bacteria; stromatolites have been in existence for the past 3.5 billion years
subduction zone	an area where the oceanic plate dives below a continental plate into the asthenosphere. Ocean trenches are the surface expression of a subduction zone
subsidence	the compaction of sediments due to the removal of fluids
surge glacier	a continental glacier that heads toward the sea at a high rate of advance

syncline	a fold in which the beds slope inward toward a common axis
tectonic activity	the formation of the Earth's crust by large-scale earth movements throughout geologic time
tephra	all clastic material from dust particles to large chunks, expelled from volcanoes during eruptions
terrane	a unique crustal segment attached to a landmass
Tethys Sea	the hypothetical mid-latitude area of the oceans separating the northern and southern continents of Gondwana and Laurasia hundreds of million years ago
tillite	a sedimentary deposit composed of glacial till
transform fault	a fracture in the Earth's crust along which lateral movement occurs. They are common features of the midocean ridges
transgression	a rise in sea level that causes flooding of the shallow edges of continental margins
tsunami	pronounced sue-NA-mee, a seismic sea wave
tuff	a rock formed of pyroclastic fragments
tundra	permanently frozen ground at high latitudes and high altitudes
unconformity	a surface marked by erosion or nondeposition
uniformitarianism	the belief that the slow processes that shape the Earth's surface have acted essentially unchanged throughout geologic time
varve	thinly laminated lake bed sediments
viscosity	the resistance of a liquid to flow
volcanic ash	fine pyroclastic material injected into the atmosphere by an erupting volcano
volcanic bomb	a solidified blob of molten rock ejected from a volcano
volcanic cone	the general term applied to any volcanic mountain with a conical shape

volcanic crater	the inverted conical depression found at the summit of most volcanoes, formed by volcanic eruption
volcano	a fissure or vent in the crust through which molten rock rises to the surface to form a mountain
volcanism	any type of volcanic activity

BIBLIOGRAPHY

THE EARTH'S CRUST

Burchfiel, B. Clark. "The Continental Crust." *Scientific American* 249 (September 1983): 130–142.

Francheteau, Jean. "The Oceanic Crust." *Scientific American* 249 (September 1983): 114–129.

Jones, Richard C. "Unraveling Origins, the Archean." *Earth Science* 42 (Winter 1989): 20–22.

Jordan, Thomas H. "The Deep Structure of the Continents." *Scientific American* 240 (January 1979): 92–107.

Kerr, Richard A. "The Continental Plates Are Getting Thicker." *Science* 232 (May 23, 1986): 933–934.

Kozlovsky, Ye. A. "The World's Deepest Well." *Scientific American* 251 (December 1984): 98–104.

Monastersky, Richard. "The Whole-Earth Syndrome." *Science News* 133 (June 11, 1988): 378–380.

Siever, Raymond. "The Steady State of the Earth's Crust, Atmosphere and Oceans." *Scientific American* 230 (June 1974): 72–79.

Sun, Shen-su. "Multistage Accretion and Core Formation of the Earth." *Nature* 313 (February 21, 1985): 628–629.

Weisburd, Stefi. "How Hot Is the Heart of the Earth?" *Science News* 131 (April 18, 1987): 245.

Wetherill, George W. "The Formation of the Earth from Planetesimals." *Scientific American* 224 (June 1981): 163–174.

BASEMENT ROCK

Brock, Thomas D. "Precambrian Evolution." *Nature* 288 (November 20, 1980): 214–215.

Fisher, Arthur. "Alaska Down Under?" *Popular Science* 228 (June 1988): 10–12.

Howell, David G. "Terranes." *Scientific American* 253 (November 1985): 116–125.

Jones, David L., et al. "The Growth of Western North America." *Scientific American* 247 (November 1982): 70–84.

Kerr, Richard A. "Another Movement in the Dance of the Plates." *Science* 244 (May 5, 1989): 529–530.

———. "Plate Tectonics Is the Key to the Distant Past." *Science* 234 (November 7, 1987): 670–672.

Krogstad, E. J., et al. "Plate Tectonics 2.5 Billion Years Ago: Evidence at Kolar, South India." *Science* 243 (March 10, 1989): 1,337–1,339.

Kunzig, Robert. "Birth of a Nation." *Discover* 11 (February 1990): 26–27.

———. "Horizontal History." *Discover* 10 (September 1989): 16–18.

Molnar, Peter and Paul Tapponnier. "The Collision between India and Eurasia." *Scientific American* 236 (April 1977): 30–40.

Moorbath, Stephen. "The Oldest Rocks and the Growth of Continents." *Scientific American* 236 (March 1977): 92–104.

SEDIMENTARY STRATA

Berner, Robert A., and Antonio C. Lassaga. "Modeling the Geochemical Carbon Cycle." *Scientific American* 260 (March 1989): 74–81.

Blatt, Harvey, Gerald Middleton, and Ramond Murry. *Origin of Sedimentary Rocks.* New York: Prentice-Hall, 1972.

Broecker, Wallace S. "The Ocean." *Scientific American* 249 (September 1983): 146–160.

Dietz, Robert S., and Mitchell Woodhouse. "Mediterranean Theory May Be All Wet." *Geotimes* 33 (May 1988): 4.

Friedman, Gerald M. "Slides and Slumps." *Earth Science* 41 (Fall 1988): 21–23.

Idso, Sherwood B. "Dust Storms." *Scientific American* 235 (October 1976): 108–114.

Iisu, Kenneth J. "When the Mediterranean Dried Up." *Scientific American* 227 (December 1972): 27–36.

Matthews, Robley K. *Dynamic Stratigraphy*. New York: Prentice-Hall, 1974.

Talbot, Christopher J., and Martin P. A. Jackson. "Salt Tectonics." *Scientific American* 257 (August 1987): 70–79.

Erosional Processes

Abelson, Philip H. "Climate and Water." *Science* 260 (January 27, 1989): 461.

Ambroggi, Robert P. "Water." *Scientific American* 243 (September 1980): 101–115.

Cathles, Lawrence M., III. "Scales and Effects of Fluid Flow in the Upper Crust." *Science* 248 (April 20, 1990): 323–328.

"Facing Geologic and Hydrologic Hazards." *U.S. Geological Survey Professional Paper 1240-B*. Government Printing Office, 1981.

Francis, Peter, and Stephen Self. "Collapsing Volcanoes." *Scientific American* 256 (June 1987): 91–97.

Gibbons, Boyd. "Do We Treat Our Soil Like Dirt?" *National Geographic* 166 (September 1984): 353–388.

Monastersky, Richard. "You Just Can't Wear Them Down." *Science News* 132 (November 7, 1987): 301.

———. "Soil May Signal Imminent Landslide." *Science News* 134 (November 12, 1988): 318.

———. "Spotting Erosion from Space." *Science News* 136 (July 22, 1989): 61.

Type Sections

Badash, Lawrence. "The Age-of-the-Earth Debate." *Scientific American* 261 (August 1989): 90–96.

Goetz, Alexander F. H. "Geologic Remote Sensing." *Science* 211 (February 20, 1981): 781–790.

Maslowski, Andy. "Eyes on the Earth." *Astronomy* 14 (August 1986): 9–10.

Mintz, Leigh W. *Historical Geology*. Columbus, OH: Charles E. Merrill, 1972.

Monastersky, Richard. "Ancient Ocean Upheaval Marks the Spot." *Science News* 136 (July 22, 1989): 61.

Simon, Cheryl. "In With the Older." *Science News* 123 (May 7, 1983): 300–301.

———. "The Great Earth Debate." *Science News* 121 (March 13, 1982): 178–179.

Stokes, W. Lee. *Essentials of Earth History*. New York: Prentice Hall, 1982.

FOSSIL BEDS

Averett, Walter R. "Fertile Fossil Field." *Earth Science* 41 (Spring 1988): 16–18.

Burgin, Tini. "The Fossils of Monte San Giorgio." *Scientific American* 260 (June 1989): 74–81.

Fisher, Louise J. "Finding Fossils." *Earth Science* 41 (Summer 1988): 20–22.

Hannibal, Joseph T. "Quarries Yield Rare Paleozoic Fossils." *Geotimes* 33 (July 1988): 10–13.

Jeffery, David. "Fossils: Annals of Life Written in Rock." *National Geographic* 168 (August 1985): 182–191.

Monastersky, Richard. "Fossils Push Back Origin of Land Animals." *Science News* 138 (November 10, 1990): 292.

———. "Supersoil." *Science News* 136 (December 9, 1989): 376–377.

Morris, S. Conway. "Burgess Shale Faunas and the Cambrian Explosion." *Scientific American* 246 (October 20, 1989): 339–345.

Morris, S. Conway, and H. B. Whittington. "The Animals of the Burgess Shale." *Scientific American* 241 (July 1979): 122–133.

Mossman, David J., and William A. S. Sarjeant. "The Footprints of Extinct Animals." *Scientific American* 248 (January 1983): 75–85.

Stolzenburg, William. "When Life Got Hard." *Science News* 138 (August 25, 1990): 120–123.

FOLDING AND FAULTING

Bird, Peter. "Formation of the Rocky Mountains, Western United States: A Continuum Computer Model." *Science* 239 (March 25, 1988): 1501–1507.

Cook, Frederick A., Larry D. Brown, and Jack E. Oliver. "The Southern Appalachians and the Growth of Continents." *Scientific American* 243 (October 1980): 156–168.

Frolich, Cliff. "Deep Earthquakes." *Scientific American* 260 (January 1989): 48–55.

Gore, Rick. "Our Restless Planet Earth." *National Geographic* 168 (August 1985): 142–179.

James, David E. "The Evolution of the Andes." *Scientific American* 229 (August 1973): 63–69.

Johnston, A. C. "A Major Earthquake Zone on the Mississippi." *Scientific American* 246 (April 1982): 60–68.

Johnston, A. C., and Lisa R. Kanter. "Earthquakes in Stable Continental Crust." *Scientific American* 262 (March 1990): 68–75.

Jordan, Thomas H., and J. Bernard Minster. "Measuring Crustal Deformation in the American West." *Scientific American* 259 (August 1988): 48–55.

Kerr, Richard A. "Delving into Faults and Earthquake Behavior." *Science* 235 (January 9, 1987): 165–166.

Mohnar, Peter. "The Structure of Mountain Ranges." *Scientific American* 255 (July 1986): 70–79.

Stein, Ross S., and Robert S. Yeats. "Hidden Earthquakes." *Scientific American* 260 (June 1989): 48–57.

IGNEOUS ACTIVITY

Bonatti, Enrico. "The Origin of Metal Deposits in the Oceanic Lithosphere." *Scientific American* 238 (February 1978): 54–61.

Cox, Keith G. "Kimberlite Pipes." *Scientific American* 238 (April 1978): 120–132.

Hekinian, Roger. "Undersea Volcanoes." *Scientific American* 251 (July 1984): 46–55.

Kittleman, Laurence R. "Tephra." *Scientific American* 241 (December 1979): 160–170.

Macdonald, Kenneth C., and Paul J. Fox. "The Mid-Ocean Ridge." *Scientific American* 262 (June 1990): 72–79.

Oxburgh, E. Ronald, and R. Keith O'Nions. "Helium Loss, Tectonics, and the Terrestrial Heat Budget." *Science* 237 (September 25, 1987): 1583–1587.

Peck, Dallas L., Thomas L. Wright, and Robert W. Decker. "The Lava Lakes of Kilauea." *Scientific American* 241 (October 1979): 114–128.

Rampino, Michael R., and Richard B. Strothers. "Flood Basalt Volcanism during the Past 250 Million Years." *Science* 241 (August 5, 1988): 663–667.

Rona, Peter A. "Mineral Deposits from Sea-Floor Hot Springs." *Scientific American* 254 (January 1986): 84–91.

Vink, Gregory E., W. Jason Morgan, and Peter R. Vogt. "The Earth's Hot Spots." *Scientific American* 252 (April 1985): 50–57.

White, Robert S., and Dan P. McKenzie. "Volcanism at Rifts." *Scientific American* 261 (July 1989): 62–71.

Wickelgren, Ingrid. "Simmering Planet." *Discover* 11 (July 1990): 73–75.

Glacial Terrain

Barnes-Svarney, Patricia. "Beyond the Ice Sheet." *Earth Science* 39 (Summer 1986): 18–19.

Bowen, D. Q. "Antarctic Ice Surges and Theories of Glaciation." *Nature* 283 (February 14, 1980): 619–620.

Broeker, Wallace S., and George H. Denton. "What Drives Glacial Cycles." *Scientific American* 262 (January 1990): 49–56.

Kerr, Richard A. "Milankovitch Climate Cycles through the Ages." *Science* 235 (February 27, 1987): 973–974.

Krantz, William B., Kevin J. Gleason, and Nelson Cain. "Patterned Ground." *Scientific American* 259 (December 1988): 68–76.

Mathews, Samuel W. "Ice on the World." *National Geographic* 171 (January 1987): 84–103.

Mollenhauer, Erik, and George Bartunek. "Glacier on the Move." *Earth Science* 41 (Spring 1988): 21–24.

Monastersky, Richard. "Hills Point to Catastrophic Ice Age Floods." *Science News* 136 (September 30, 1989): 213.

Moran, Joseph M., Ronald D. Stieglitz, and Donn P. Quigley. "Glacial Geology." *Earth Science* 41 (Winter 1988): 16–18.

Weisburd, Stefi. "Halos of Stone." *Science News* 127 (January 19, 1985): 42–44.

Unique Structures

Aydin, Atilla, and James M. DeGraff. "Evolution of Polygonal Fracture Patterns in Lava Flows." *Science* 239 (January 29, 1988): 471–475.

Bolton, David W. "Underground Frontiers." *Earth Science* 40 (Summer 1987): 16–18.

Bonatti, Enrico. "The Rifting of Continents." *Scientific American* 256 (March 1987): 97–103.

Daniel, Glyn. "Megalithic Monuments." *Scientific American* 243 (July 1980): 78–90.

Francis, Peter. "Giant Volcanic Calderas." *Scientific American* 248 (June 1983): 60–70.

Grieve, Richard A. F. "Impact Cratering on the Earth." *Scientific American* 262 (April 1990): 66–73.

Hansen, Michael C. "Ohio Natural Bridges." *Earth Science* 41 (Winter 1988): 10–12.

Kerr, Richard A. "Is the San Andreas Weak at Heart." *Science* 236 (April 24, 1987): 388–389.

BIBLIOGRAPHY

Macdonald, Ken C., and Bruce P. Luyendky. "The Crest of the East Pacific Rise." *Scientific American* 224 (May 1981): 100–116.

Sharpton, Virgil L. "Glasses Sharpen Impact Views." *Geotimes* 33 (June 1988): 10–11.

Weisburd, Stefi. "Sensing the Voids Underground." *Science News* 130 (November 22, 1986): 329.

INDEX

A

B

C

T

U

V

W

X

Y

Z